The Greening of Pentagon Brownfields

The Greening of Pentagon Brownfields

Using Environmental Discourse to Redevelop Former Military Bases

KENNETH N. HANSEN

LEXINGTON BOOKS
Lanham • Boulder • New York • Toronto • Oxford

LEXINGTON BOOKS

Published in the United States of America
by Lexington Books
An imprint of The Rowman & Littlefield Publishing Group, Inc.
4501 Forbes Boulevard, Suite 200, Lanham, Maryland 20706

PO Box 317
Oxford
OX2 9RU, UK

British Library Cataloguing in Publication Information Available

Library of Congress Cataloging-in-Publication Data

Hansen, Kenneth N., 1967-
The greening of Pentagon brownfields : using environmental discourse to redevelop former military bases / Kenneth N. Hansen.
p. cm.
Includes bibliographical references and index.
ISBN 978-0-7391-0539-9
1. Military base conversion—Environmental aspects—United States. 2. Hazardous waste site remediation—United States. 3. Environmental policy—United States. I. Title.
TD195.A75H35 2004
363.73'2'0973—dc22 2004003388

Printed in the United States of America

∞™ The paper used in this publication meets the minimum requirements of American National Standard for Information Sciences—Permanence of Paper for Printed Library Materials, ANSI/NISO Z39.48–1992.

Table of Contents

Acknowledgments

The late Bob Bullock, who served as lieutenant governor of Texas for many years, once told me that all successful people have help on their way to the top, no matter how much they may think they made it on their own. Keeping Bob's words in mind, though I have done a great deal on my own, I do not want to forget those many people who have helped me on the path to becoming a better person and a successful professional. Inevitably, I am going to leave a number of people out, but I hope they do not take it personally.

My many friends and family have given me a great deal of emotional and financial support throughout the years. They include, but are certainly not limited to, my parents, Norman and Helena Hansen; my late grandmothers Erna Clarke Bakker and Doris Hansen; my "sibs," Michael Hansen, Norene, Gregg and Sierra Gill; uncles and aunts John and Joy Taylor, and Dwayne and Darlene Hansen (for letting me stay with them in Arizona on my many road trips between Texas and California); my best friends, Don and Patti; and my namesake, Kenny Peterson. I owe a debt of gratitude also to J. J. St. Marie, for listening to my constant ranting, Jenn Flaherty, for always being there and for staying alive when we thought you had left us for good, the Stumm clan, Jeana Collignon-Comish, the Diaz clan, Steve Ellers, Dan Krejci, Judy Fai-Podlipnik, Jeff Ryan, Bill Schreckhise and Janine Parry, for letting me stay in the "Ken Hansen suite" during periods of intense insobriety, and Steven Neuse, for recruiting me away from the Crawfish Republic. These are but a few of the many people who have touched my life in some way for the better.

This book, and my academic career, would not have been possible without the guidance of Mark Somma, now at California State University-Fresno, Charles J. Fox at Texas Tech University, and Ted Stumm (colonel, U.S. Air Force, retired) at the University of North Florida. Little did we know that what started as a small research project for the city of Lubbock's base retention effort would result in an entire research agenda! Tracy Skopek, now at the University of Wyoming, was a big help on data collection for the subsequent Texas Tech University Military Base Reuse and Community Development project. I also owe many thanks to Donna Barnes and Evelyn Nobles in the Texas Tech Department of Political Science and the Center for Public Service. They are living examples of the phrase "overworked and underpaid," but they are certainly appreciated!

This endeavor would not have been completed were it not for the support of my colleagues in the Political Science Department at the University of Arkansas. My department chair, Todd Shields, was more than generous with departmental resources and allowing me the time to write. His comments on the book proposal and his encouragement during my periods of distraction and writer's block were invaluable. J. Brinck Kerr III and the Center for the Study of Representation supported me with numerous travel grants and critical feedback during faculty research presentations. The rest of the faculty, including former U.S. senator David Pryor, were likewise supportive and encouraging. Of my graduate assistants, Michelle Berg-Vogel deserves special thanks for helping me circulate the proposal to numerous publishers, while Gaby Jabbour was helpful during the editing phase. Our department staff, Linda Gardner and Laura Bowers, helped keep me organized—to the extent that it's even possible. Laura also gets a big thank you for helping me with all my computer problems.

At Southeastern Louisiana University in the Crawfish Republic I would like to thank Jeff Bell for reading an early version of chapter II, Bill Robison for reading a draft of the book proposal, and Michael Kurtz for the research grant and approval to administer the survey. But the one who deserves the most thanks is Dawn A. Diaz, who was my research assistant on the base conversion survey, and who went above and beyond the call of duty when I was diagnosed with cancer and broke my leg. It is very difficult to work when one is in poor health, but she kept the ball rolling for me during the hard times.

I would like to thank my colleagues and department staff in Political Science and Public Administration at the University of North Florida for their support of the Military Adjustment Study Group during my time in Jacksonville, and for their hospitality on my frequent return visits.

The base conversion community was extremely helpful over the years with regard to linking theory and practice. I would like to thank those anonymous executive directors and staff who responded to the base conversion survey. I would also like to thank the National Association of Installation Developers consultants, active-duty and retired military, federal, state, and local officials, whose conversations helped cast light on the complexities of military base redevelopment. These include, in no particular order, Bobbi Morgan, Jim Meyer, Brad Arvin, Derrick Curtis, Olan Long, Jim Steed, Jim Bertram, Duane Lavery, Kenny Maines, Jerry Millsaps (sergeant, U.S. Air Force), Herb McCarthy (captain, U.S. Navy, retired), Mike Weinstein, Jim Fuller (R-Jacksonville, FL), Jim Overton, Clarence Juhl from NATO (admiral, U.S. Navy, retired), Rolando Bono, Jesse Lopez, David Dimaline, Paul Roberson (general, U.S. Air Force, retired), Dick Pautz, Frank Gray, Kevin Walker, Ron Weeger, Jeanette Hook, Jackie Ellis, Dean Kruitof, Jerry Canfield, Noel Bennett, Larry Evans, S. McDonald, and D. Jones.

My editor, Jason Hallman, assistant acquisitions editor Rebekka Brooks, editorial assistant Rebecca Olson, and the reviewers for Lexington Books get my undying gratitude for publishing my work and making it presentable. Any errors and omissions are, of course, the responsibility of the author.

And last, but not least, I cannot forget my students, who have taught me so much: Thank you for encouraging my behavior!

Acronyms

AFB	Air Force Base
AFBCA	Air Force Base Conversion Agency
ALC	Air Logistics Center
AETC	Air Education Training Command (formerly ATC)
BCT	BRAC Clean-up Team
BEC	Base Environmental Coordinator
BLM	Bureau of Land Management
BRAC	Base Realignment and Closure
BTC	Base Transition Coordinator
CBO	Congressional Budget Office
CERCLA	Comprehensive Environmental Response Compensation and Liability Act of 1980, "Superfund"
DERP	Defense Environmental Restoration Program
DOD	Department of Defense
DOE	Department of Energy
DOI	Department of the Interior
EDA	Economic Development Agency
EDC	Economic Development Conveyance
EIS	Environmental Impact Statement
EPA	Environmental Protection Agency
ETRPA	El Toro Reuse Planning Authority
FAA	Federal Aviation Administration
FDEP	Florida Department of Environmental Protection
GAO	General Accounting Office
HUD	Housing and Urban Development
IRB	Installation Restoration Board
IRP	Installation Restoration Program
LRA	Local Reuse Authority
LRRA	Lubbock-Reese Redevelopment Authority
MCAS	Marine Corps Air Station

NAID	National Association of Installation Developers
NAS	Naval Air Station
NEPA	National Environmental Protection Act of 1969
NPL	National Priorities List
OEA	Office of Economic Adjustment
PBC	Public Benefit Conveyance
PCB	Polychlorinated Biphenals
PDA	Pease Development Authority
QDR	Quadrennial Defense Review
RAB	Restoration Advisory Board
RCRA	Resource Conservation and Recovery Act of 1976
SAC	Strategic Air Command
SARA	Superfund Amendments and Reauthorization Act of 1986
TAC	Tactical Air Command
TCE	Trichloroethylene
TNRCC	Texas Natural Resource Conservation Commission
UPT	Undergraduate Pilot Training
USAF	United States Air Force
USMC	United States Marine Corps
USN	United States Navy

Chapter I

Introduction

Our story begins in the "cultural mecca" of Lubbock, Texas. In April 1997, the last T-1A Jayhawk trainer aircraft vacated Reese Air Force Base (AFB). With it went the millions of dollars and hundreds of jobs that had benefited the local economy on an annual basis before operations on the base ceased. Members of the community and political leaders had been faced with the question, "what do we do next?" for the previous two years. Of additional concern was the groundwater contamination caused by a plume of chemical cleaning solvent that was rapidly spreading off of the base property. This threatened the health of residents and the prosperity of businesses adjacent to the installation, as well as the redevelopment potential of the facility. Parts of the base that had important redevelopment potential, such as the golf course and the lake at the picnic area, which are located just north of the aptly—if inadvertently—named Brownfield Highway, had to be parceled out for the rest of the base to be used. Other former bases in Texas, such as Bergstrom AFB in Austin and Carswell AFB in Fort Worth, experienced much more extensive pollution problems and were forced to implement extreme measures in order to complete their redevelopment plans.

Southern Orange County, California, officials are faced with a similar dilemma as they confront the closure of the El Toro Marine Corps Air Station (MCAS). Although they have had more time to deal with the problem, these officials face a scale of difficulty somewhat greater than that faced by Lubbock. El Toro employed substantially more personnel, and the impact from its drawdown represents a potentially larger problem for long-term growth of the regional economy than does the closure of Reese AFB. However, local leaders and citizens have not been able to reach a consensus regarding redevelopment plans. Of particular concern are noise and air pollution levels that will result if the base is converted to an international airport. To date, the county government has not been able to answer questions concerning the overall quality of life for its citizens if its redevelopment plans go forward.

The consolidated city/county government of Jacksonville, Florida, has had similar NIMBY (Not In My Backyard) problems with its transition of Naval Air

Station (NAS) Cecil Field, located just west of the main urban population. Cecil Field was sacrificed to protect Jacksonville's two other main military installations, the Mayport Naval Station, which is home port to several battle groups, and Naval Air Station Jacksonville (NAS Jax), an air maintenance depot which employs thousands of civilian workers. Cecil Field, however, is of limited redevelopment potential compared to the other two facilities, and is home to major pollution sites. The first redevelopment plan involved building a jail on the former base, but it was shot down by angry, mobilized citizens. Much of the facility has since reverted to the local port authority (Jaxport), which is attempting to integrate it with its intermodal shipping enterprise. What all three of these cases, along with many others, have in common, is that their redevelopment potential and timetables may be adversely affected by pollution left behind by the military or from the proposed redevelopment solutions themselves.

Of the ninety-seven major military installations (and eighteen minor ones) which have been ordered to close or realign since 1988, there are many examples of success, along with several failures, and many of which seem to fall somewhere in between. State and local governments in the United States are increasingly being called upon to deal with the problems associated with defense downsizing and environmental damage caused by the military. What is unknown to many people is that the U.S. government pollutes as much as any large corporation, maybe more. Of the 116 federal facilities on the EPA's "hit list" in the early 1990s of the most polluted sites in the United States, 95 were military installations (Steppick 1994). The military is responsible for cleaning up the mess it made, which will cost tens of billions of taxpayer dollars, before property can be transferred for civilian use (CBO 1996). These events pose a multitude of questions regarding the role of state and local governments and the nature of domestic defense policy implementation. At stake are thousands of jobs and millions of dollars in investment opportunities, as well as the environmental health and quality of life in many parts of the country. This work explores how states and localities have increased their institutional capacities to deal with the unanticipated consequences of federal downsizing and pollution at military bases in the 1990s.

Since the 1970s, the United States has experimented with what has come to be called devolution. Devolution involves transferring functions of government that were previously implemented by the national government to subnational (state and local) governments (Hedge and Scicchitano 1992). Richard Nixon began the process of increasing state institutional capacity through revenue sharing, but ironically this was discontinued under Ronald Reagan's New Federalism. During the Reagan years, devolution was advanced through the appointment of officials hostile to federal regulation and amenable to increasing state powers, and also through reducing federal agencies' capacities to regulate (Hedge and Scicchitano 1992). By the early 1990s, Congress had further increased the responsibilities of subnational governments through the unpopular practice of unfunded mandates, which were largely discontinued in 1995, though "partially funded" mandates still exist. This is the practice whereby the federal government funds a new program for two or three years, after which time states and localities must pick up the funding.

The Bill Clinton years saw further reduction in the national capacity through the initiation of reinventing government through the National Performance Review. This reduced the number of federal employees by 12 percent, mostly through attrition. What is often left out is that much of the reduction in federal personnel came from Pentagon downsizing. From 1988 to 1995, the Defense Department lost 30 percent of its personnel and 15 percent of its domestic base infrastructure. Defense downsizing is not considered a direct example of devolution.[1] However, its effects are much the same because subnational governments have to deal with its effects without the benefit of increased institutional capacities. The result is that states, being the newest of environmental regulators (which in fact is an example of devolution according to Hedge and Scicchitano 1992), must deal with polluted facilities, while local governments deal with potentially devastating unemployment and recession through economic development. The question is, how do they do this?

The Base Closure Conundrum

Over the years, many local governments have relied on military installations as a source of revenue and employment opportunities for their citizens. The seemingly permanent and predictable economic benefits were often taken for granted, until the federal government began downsizing the base infrastructure of the Department of Defense following the end of the Cold War. In the early 1990s, many communities put forth a great deal of effort to save their bases from the budget-cutting ax of the BRAC (Base Realignment and Closure) Commission, but gave little thought to what they would do if they actually lost their installations. Due to the emphasis on base retention, many localities were not sufficiently prepared for the base conversion process once final closure decisions had been made and approved (Hansen et al. 1997).

Although closing a military base may have crippling economic effects on a community, this is by no means the only possible outcome. Closed installations can represent considerable economic development opportunities. However, due to the expedience of depending on the federal dollars associated with operational bases during the Cold War, little effort was made by localities to realize the potential of economic diversification. Reasons for this include convenience, the difficulty of mobilizing diverse interests within a community, and the risk of significant amounts of investment capital. Pentagon downsizing has forced local governments to adapt in the immediate post-Cold War era by delegating to them the task of base conversion. Most local governments have little experience with such activities, which demand creative methods of planning and implementation.

As if this were not enough, the facilities inherited by local governments from the Pentagon are often polluted. While the DOD now admits—indeed they are liable under federal law—to having contaminated most of the closing military installations during the Cold War, the subsequent cleanup can still potentially delay redevelopment for many years. Most local governments do not have the resources to clean up contaminated installations. The Pentagon is obligated for the cost, which alleviates the largest burden, but other difficulties still remain. Each base is unique, with many of them having had several missions with correspondingly different environmental consequences.

The "peace dividend" was supposed to come with the end of the Cold War and the downsizing of the domestic defense infrastructure. Due to the cost of environmental problems, the proverbial peace dividend will not be realized. The Congressional Budget Office estimated in 1996 that the federal government is obligated to pay as much as $6.6 billion to clean up pollution caused by the Department of Defense at eighty-four of the ninety-seven major closed military installations (CBO 1996). That, along with the practice of giving base property to communities through Economic Development Conveyance (EDC), makes defense downsizing unprofitable for the U.S. government, and hence the taxpayers. I suspect the eventual cost is probably greater, but we may never know the full amount, since money for cleanup and redevelopment comes from a variety of different budgetary sources, thus obscuring the total costs.

This work provides a critique of the closure process and then uses it as a starting point for an analysis of base transitions. Key intervening variables in this process include dealing with environmental problems caused by the military during its lengthy residence at closing and former bases. This often necessitates state involvement, as traditionally localities are seen as being too small to deal with large-scale environmental difficulties (Mazmanian and Sabatier 1983). Nevertheless, localities and their states often cooperate in clean-up efforts at former military bases. In addition to polluted installations, subnational governments are also concerned with contamination resulting from the construction and dismantling of nuclear weapons.

The theoretical perspective of this book is based on current views in the field of public administration and environmental policy. There are several reasons for this. The first is that no single theory yet exists which can adequately explain domestic defense downsizing. The second is that current efforts to do so come primarily from the field of American politics, and hence tend to deemphasize the administrative aspects of this kind of policy. The third is that some of the most ground-breaking theory in political science is coming from the public administration and policy experts. Other fields which might logically be concerned with defense downsizing, such as international relations, have by and large ignored the topic altogether and contributed little useful theory which might be applied to enhancing our understanding.

Recent works on base closure and conversion emphasize politics, because that is where the action is. The process was designed to be ostensibly nonpolitical, but by most accounts, no one was fooled by the pretense (Koven 1992; Mayer 1995; Sorenson 1998). The danger of placing too much of the focus on the politics of base closure is that the crucial implementation processes of base redevelopment and environmental cleanup may be forgotten. In the long term, these are far more important than questions regarding whose congressional districts lost their bases. However, politics is just as important in policy implementation as in anything else, and as such is important for this study as well.

The State of the Study of Defense Policy

In a ground-breaking *Public Administration Review* article, Ken Mayer and Anne Khademian (1996) argue that defense policy and American political institutions and processes have been ignored for too long—essentially since the 1960s. This is probably due to the assumption that defense policy is not visible outside of the "nonpolitical" bureaucratic model of administration. Much of the international relations literature assumes realist, rational views of defense policy, which reduce the units of analysis to a single actor in the form of the nation-state. This is true of the argument put forth by Randall Ripley and Grace Franklin (1991), who argue that defense policies are often redistributive (based on Ted Lowi's [1969] policy typology) and that they are made behind closed doors. This literature (with the exception of Huntington 1961) leaves out the multiplicity of actors in this process, as well as its negotiative characteristics.

This view of defense policy is reinforced by elected politicians, who often declare that "politics stops at the water's edge." The fact of the matter is U.S. defense policy involves a great deal of partisan politics, despite appearances of bipartisanship. The ostensibly nonpolitical BRAC process of the early 1990s was in fact very political (Koven 1992; Mayer 1995; Sorenson 1998), as are the resulting economic development decisions involved with base conversion. The perspective of this book is in accord with the newly rediscovered political view of defense policy and seeks to add to it by analyzing the politics involved with the implementation of strategic transition plans. This involves looking at variables that measure both processes and outcomes.

Two works in particular influenced the theoretical structure and design of this book. One is them is older, the other fairly recent. The first book that influenced this one was Graham Allison's *Essence of Decision* (1971). In it, Allison describes a single event, the Cuban Missile Crisis, with three competing theoretical models.

Allison's three perspectives are the rational actor, organizational behavior, and bureaucratic policy models. The rational actor model assumes that individuals and organizations will pursue their own self-interests. Game theory is often used to explain rationalism and has taken something of a stranglehold on the way international relations and American politics are taught. Organizational behavior explores the structures and process involved in public agencies and is often taught as a core course in public administration programs. It used to be assumed that bureaucratic politics were separate from electoral politics and that decisions were made at the top of agencies and filtered down to subordinates. This Wilsonian view has since been discarded in favor of more bottom-up, multichanneled organizational communication. Since Allison's time, rationalism has largely fallen out of favor in public administration, while organizational development has grown in popularity from its nascent form as portrayed in Allison's work. The study of bureaucratic politics is, of course, alive and well.

While I was continuing to research base transitions and contemplating the structure of this book, David Sorenson of the Air War College in Alabama published *Shutting Down the Cold War: The Politics of Military Base Closure* (1998). This is by far the best work on the topic to date. Sorenson also utilizes three models in his work, but since he comes from the perspective of American politics, his models are somewhat different from Allison's, despite obvious parallels. They are constituency service, policy learning, and bureaucratic politics. The constituency service model is essentially a rationalistic view of congressional behavior and attempts to explain base closure politics via pork barrel. The policy learning model explains how individuals, institutions (in particular cities and defense agencies), and politicians learned to manipulate the closure process over time. Sorenson provides evidence for this by writing a chapter on each round in the BRAC process since the original BRAC experiment in 1988 and exploring changes in strategy. This perspective is also more rationalistic than my environmental policy approach, which is more normatively value-oriented. The third model, bureaucratic politics, is very similar to Allison's and describes interagency competition among the respective branches of the service and the disagreements between the president, Congress, and the military.

Because my perspective is that of public administration rather than American politics or international relations, my two models are different—the purpose being to provide a contrast as well as a broader view of domestic defense policy than has to date been articulated by current works. Or in other words, I would like to build on the foundation laid by such defense policy writers as Allison (1971), Steven Koven (1992), Andrew Glassberg (1995), Kenneth Mayer (1995 and 1996 with Anne Khademian), and Sorenson (1998). I also have another aim in mind. I would like to infuse the theory of domestic defense policy with a broader perspective—administrative discourse—which reconciles our need for protection from foreign foes, local job creation, and a clean environment. When resolved, these seemingly incompatible goals can add up to a better quality of life for all Americans. In addition, there is a practical reason. I hope that by linking theory with practice that positive outcomes will be realized. Jobs will be created, people will make money, cities will expand their tax bases and provide more and better services, the environment will be cleaner, the quality of life for all Americans will increase, and government at all levels will work better for its citizens.

Previous works on the subject of military transition and pollution focus on descriptive processes and prescriptions as opposed to in-depth analysis. This is because at the time there was a lack of basic information. A coauthored piece which appeared in *The American Review of Public Administration* (Hansen et al. 1997), a coauthored book chapter (Hansen and Skopek 1997) and my unpublished Ph.D. dissertation (Hansen 1996) relied upon comparative case studies to describe local responses to the shock of base closure. By and large these were descriptive, but not terribly analytical. This work hopes to address that problem. Other works on the subject (Glassberg 1995 and 1996; Kirby 1992; Koven 1992) are also similarly descriptive and informative, if somewhat more analytical, but a certain amount of time has passed since much of the research in these works was done. What this means is that we now have the greater benefit of hindsight and an additional amount of data to be evaluated.

The Argument

A recent article in the magazine *Governing* illustrates the point of view cities have when environmental degradation stands in the way of economic development (Fulton and Shigley 2000). Rather than treating contaminated sites as problems to be shunned or shut down, many cities treat them as redevelopment opportunities. Why? According to most current theories of urban development, cities are in competition with one another, have normative views of how they perceive themselves and how they want others to perceive them, and are in hot pursuit of expanding their tax bases (Pagano and Bowman 1995; Peterson 1981; Stone and Sanders 1987). This book argues and provides empirical evidence through a triangulation of methods that environmental variables are among the key intervening variables in the base redevelopment process, and by extension, in similar forms of economic development situations.

These variables include the different types of pollution, the cost of cleanup, and the different cleanup solutions utilized by coordinating federal, state, and local agencies. Along the way, environmental impact statements, grant writing, strategic planning, and economic incentives play important—and statistically significant—roles. The processes used are compared with the resulting policy outcomes—jobs created and income generated by base redevelopment. The findings substantiate the main argument that unless environmental degradation is accounted for and dealt with, it will in fact delay or even thwart military base conversion to civilian use. This is not in the interest of state and local governments, nor in the interests of local communities and individuals adversely affected by defense downsizing. Henceforth, former (and one would assume, current) installations must be green.

Developmental Discourse

The theoretical perspective often used in previous studies on base conversion is discourse theory. Administrative discourse is also the underlying perspective of this work. A discourse is a high-level discussion of "what to do next." A base closure can have a sudden impact on the psyche of a community and change how they view themselves, known as a "perceptual orbit" (see Pagano and Bowman 1995). What was once seen as unthinkable—losing jobs and income that were once thought to be untouchable—has happened.

The question then becomes how to deal with the problem. The usual practice, which is facilitated by the federal government, is for community leaders to gather input from citizens and formulate a strategic plan for redevelopment. Rather than concocting developmental policy in a closed "iron triangle" or smoke-filled back room, discussions are designed to be bottom-up and open. The use of discourse is designed to make public administration less hierarchical (Fox and Miller 1995; Wamsley and Wolf 1996).

Charles Fox and Hugh Miller (1995) in particular discount the prevailing idea of the "representative democratic accountability feedback loop" of policy-making. The feedback loop assumes that policy-making is automatic and mechanistic, that people elect public officials, who then implement their preferred policy choices. Politics assumedly ends with the legislative outcome, after which the professional expertise of the bureaucrat takes over to make sure it is carried out. In fact, even in a closed policy-making system, there are many more actors involved. The point of discourse for Fox and Miller is to involve more people from different institutional affiliations. Discursive phenomena are becoming more common, given the greater number of interlocal agreements designed to deal with specific issues such as urban sprawl and economic development. Examples include regional councils of governments, port authorities and utilities, and special regulatory districts.

Gary Wamsley and James Wolf (1996) argue that discourse should be used to make government more democratic. For them, public administration needs to be "refounded" rather than reinvented. In an effort to create greater responsiveness and legitimacy in government, public managers need to become a greater part of the public debate and encourage greater citizen participation in public endeavors. Their disciples David Farmer (1995), Cam Stivers (1996) and O. C. McSwite (1996) advocate more involvement of previously excluded groups. The level of government that is closest to the people is of course local government, and it is at this level that we have seen more efforts at citizen involvement.

Discourse is used in this book in three ways. The first is as a critique of environmental and economic development policies. Discourse theory is the lens through which the two views of environmental policy are examined. It is also the tool used to evaluate the planning and redevelopment processes utilized in base conversions. Discourse is also used as a metatheory, to help reconcile views that are often seen as being in opposition to one another, such as militarism and environmentalism.

Fox and Miller (1995) utilize four "warrants" for discourse, which are based on social and conversational norms. Those who do not behave in a civilized manner are supposed to be excluded from the conversation, since they have nothing of substance to contribute.

There are several criteria with which Fox and Miller assess the authenticity of discourse. These include sincerity, situation-regarding intentionality, willing attention, and substantive contribution (Fox and Miller 1995, 129-59). Sincerity is necessary to build bridges of trust between those who would participate in the discourse. Situation-regarding intentionality "assures that the discourse will be *about* something, about contextually situated activities" (1995, 123). One's agenda or intentions are designed in relation to the particular context in which the discussion is to take place. Willing attention refers to the active participation in the discourse. Not only must one maintain an interest in the dialogue, but one must also join in from time to time and listen critically and attentively to the arguments of others. Substantive contribution includes providing expertise and ideas or maintaining the level (authenticity) of discourse by policing the conversation of one's peers (1995, 129-59).

Who then has a right, or warrant, to participate in the discourse? In an effort to be as democratic as possible, Fox and Miller advocate the involvement of anyone whose intentions are genuine, as long as they are willing to participate with others as equals in the pursuit of the public interest. "Communication requires equal participants. Unequal communication is oxymoronic; talk between unequals is either command or acquiescence" (1995, 116). Sincerity is important in these situations to build trust between participants who, at least initially, may know very little about one another. The participants must also be willingly engaged in the process, providing substantive contributions in accordance with the overall agenda, which is negotiated among the members of the group.

Discourse theory assumes that language used to dominate, exclude, reduce, lie, or misrepresent should be identified for what it is nondiscursive—and henceforth be exposed and eliminated. Participants should "police the discourse" as a way to maintain a certain level of substantive discussion. This is intended to help prevent self-interested tendencies from taking over the debate and forcing their agenda upon others.

A discussion of three empirical referents helps to explain how discourse can help deal with questions of economic development and environmental cleanup at closed military bases. The referents are inclusion, self-regulation, and policy outputs, and are designed to complement Fox and Miller's warrants (Hansen 1998). The first, inclusion, involves who is invited to participate in the discussion. Presumably, the more actors involved in the discussion, the less it is likely to be controlled by institutional or elite interests, and hence be more authentic. The second referent involves how the discussion is regulated. If the norms, rules, and procedures which structure the discussion are predetermined by policy elites, then the outcomes are likely to be the result of a top-down, unidirectional, and undemocratic nature—and hence antidiscursive. If the discussion is self-regulated by its participants, then it is closer to the ideal. The third referent, policy outputs, involves whether the discussion results in action being taken in pursuit of problem solving. A lack of substantive policy outputs indicates that something is wrong with the discourse. Policy outputs that result from an inclusive, self-regulated discussion, on the other hand, are likely to illustrate the benefits of discursive administration. These three empirical referents are designed to provide a more systematic measure of discourse (Hansen 1998). Briefly, here is how each empirical referent is utilized in this study.

Inclusion

Three criteria can be used to determine the nature of "inclusion." They are coalition building, the acceptance of outsiders, and community outreach activities. Local governments often form coalitions with discursive characteristics to gain support and legitimacy for stated economic development initiatives. Community development literature indicates that these coalitions are representative of the community at large. They incorporate various political factions, racial groups, business and grass-roots interests. Local governments provide the leadership necessary to coagulate these various interests into a workable unit (Stone and Sanders 1987; Sonenshein 1992; Pagano and Bowman 1995).

Another manifestation of inclusion is the acceptance of outsiders. Paul Sabatier (1988) suggests this as a measure for his advocacy coalition theory, although for him the outsiders are media elites, interest groups, and policy analysts. For discourse, outsiders might include consultants or specialized practitioners brought in for specific implementation purposes. Acceptance of outsiders is a good indication of whether discourse is inclusive with regard to solving highly technical policy dilemmas, as specialization (though this is a top-down concept) is often required to deal with the unanticipated difficulties associated with complex problems. Henry Mintzberg's *ad hocracy* idea also requires a certain level of expertise for addressing policy implementation problems, though it is generally considered to be bottom-up. The general, as opposed to specialized, nature of these structures makes it theoretically possible for many individuals to come and go as needed (Mintzberg 1994).

A third indicator of inclusion is the practice of community outreach activities. Such activities include, but are not limited to, town meetings, regular media coverage, informative surveys, or any other practices that enable or encourage public participation. Town meetings must be set up in such a way as to encourage multidirectional communication, and should be held on a regular basis at a consistently accessible location. Town meetings can be evaluated by observing the spatial arrangements of the participants and observers and the format of the discussion. A roundtable setting is most appropriate for discourse because it encourages a feeling of conversation between equals. Other types of seating arrangements, such as having lecturers on a stage or at a podium in front of an audience, are more conducive to monologic communication, as they stifle participation and the normal give-and-take of conversation by rearranging the "civic space" (Domahidy and Gilsinan 1992; Goodsell 1988).

Regular media coverage can take the form of weekly newspaper columns or television news stories, the substance of which could be evaluated through a content analysis if one were ambitious enough. The media can also keep the community informed regarding the places and times of public hearings or meetings, rather than simply reporting the results of the discussion after the fact. Informative surveys are a useful tool with which to gauge public opinion and values. They are also a useful way to increase the polity's level of knowledge, and provide a means to seek out the ideas of those who may otherwise be unable to participate in regular discussions due to high information or opportunity costs. Many base conversion efforts have done just that. Some have utilized 1-800 telephone numbers, faxes, and e-mail to provide multiple channels of communication with citizens.

Self-regulation

Self-regulation is an important aspect of discourse. If the participatory group or community is not allowed to "police the discourse," in the words of Fox and Miller, then communication is unequal, hierarchical, and nondiscursive. There can be no discursive will formation or willing engagement if it is coerced. One indicator of self-regulation is the election of leadership by problem-solving committees and subcommittees. This practice—legitimizing street-level or citizen-level decisionmaking—is often called empowerment (Peters and Savoie 1996). A criticism of this is that elections can be fixed or won through false promises. Just because a polity has elections does not mean it is democratic (Miller 1998). Another indicator is the negotiation of operating norms, rules, and procedures, which serve to add direction to the conversation. Fox and Miller refer to these as "recursive practices," which are needed to establish and maintain discursive patterns of communication. A third indication is multidirectional communication. Again, one way this can be evaluated is through the observation of spatial arrangements at meetings. If people are relegated to an audience—by seating them at other tables or in rows of chairs—they are reduced to the status of passive evaluators rather than active implementers, and are hence not allowed to take part in the discussion or its direction, that is, unless they are somehow empowered by some technological means.

Self-regulation is often difficult to get a handle on. Because it (along with discourse generally) is more organic, traditional means of social science research have a difficult time cracking it open. Policy-making is often elite driven. If local elites structure the debate and then let it go wherever it takes them, do you have discourse, or is it still dominated from the top down? With regard to questions like these, it is useful to bring in the warrants to examine whether things are as they appear or whether someone is playing the puppet master behind the scenes. The reality is probably that there are both top-down and bottom-up forces at work. Power elites are often reluctant to share power with the mob, though they may pay lip service to democracy. Breaking "good old boy" networks and traditional hierarchies in governmental institutions is much harder to do than one may realize.

Policy Outputs

Finally, policy outputs can be used to measure progress in the problem-solving process. Fox and Miller (1995) argue that discourse has to be about what to do next and that substantive contributions must be geared toward this end. I assert that "how to do it" necessarily follows from the discussion of "what to do next." However, Fox and Miller make no distinction between policy formulation and implementation. Whereas they posit that policy solutions can be negotiated through politics and that discourse is necessarily about something, I argue that discourse is not only about something, but also that to be authentic with regard to addressing complex policy problems, *it must be specifically about the implementation process*. There has to be a road map or a plan, which is the product of discourse. In base conversions, this often manifests itself in the form of strategic planning, grant writing, and environmental impact statements.

There are several steps to policy implementation that include the creation and funding of implementing agencies, the construction of rules and regulations (Cochran, et al. 1999), in addition to the daily activities of policy execution. Regardless of whether these happen sequentially, as modernists would argue, or simultaneously, as postmodernists do, they are distinct activities that comprise the implementation process, and that is the concern here. James Q. Wilson (1989) argues that every organization must have a task, mission, and the autonomy to carry out policy objectives. These criteria can be used to describe the policy output referent, because they are illustrative of specific actions resulting from discussions. If solutions are not negotiated within a certain amount of time, this may be indicative of hidden agendas, stonewalling, poor organization, or incompetence, and serve to indicate that there are problems somewhere in the process.

It should be kept in mind that while policy outputs are necessary for discourse to be effective, when taken by themselves they are not sufficient indicators, because they are not unique to discourse. Policy outputs that serve to define the measurement of implementation activity may include organizational development, funding, completed strategic plans, and their actual execution. Such activities are not exclusively discursive like inclusion or self-regulation, but they serve to illustrate the authenticity of discourse through the demonstration of tangible action. In other words, discourse is only authentic if something is done. Otherwise there is no point to having the discussion. In some instances it may be the case that a nondecision is rendered from discourse. However, with regard to environmental contamination at former military bases, this is not likely to happen. Policy outputs indicate discursive implementation because there can be no realization of "how to do it" without tangible results.

Conversely, the lack of discursive behavior can also be measured through the observation of exclusion, hierarchy, and the lack of policy outputs geared toward problem solving. Exclusion is exemplified through elite control, refusal to include outsiders, and an unwillingness to reach out to the community. Orthodoxy is most easily defined by the practices of appointing officials to a hierarchical system and unidirectional communication. A paucity of policy outputs such as organizational development, adequate funding, strategic planning, and implementation activity can be noted by their absence, or by the presence of the very antithesis of discourse—

litigation—which has been known to occur in several base conversion cases.

What, then, do the warrants and empirical referents tell us about greening former military installations so that they can be reused? If environmental variables are key barriers to economic growth, then any developmental discourse will have to deal with them. With regard to timelines and planning, environmental issues will manifest themselves as hurdles to be overcome. Whether grants and strategic implementation plans are approved may hinge on the progress of environmental impact statements and clean-up efforts. Such things can be observed both directly and indirectly.

The Outline of the Book

This book provides a series of essays that critique various facets of the subnational domestic defense downsizing processes. The methodology involves a triangulation of comparative case studies, a quantitative analysis based on results of a base conversion survey, and government documentation of both the process and outcomes. The first part of the book, which includes chapters II and III, provides an analytical framework based on two different theories of environmental policy. One explores the orthodox view of environmental policy implementation; the other looks at more current perspectives such as market incentives, intergovernmental relations (federalism and devolution), and local control (including environmental discourse and justice).

Chapter II discusses the orthodox model of environmental policy implementation, which emphasizes top-down control by federal agencies and finds it lacking. The main assumptions of this theory are that most pollution is caused by the private sector and local governments, and that the way to mediate the unintended consequences of the market and public consumption is through federal action. For this reason, new federal agencies were set up and existing agencies, including the Pentagon, expanded their focus to deal with environmental problems. Of course, part of the reason why this happened is because lower levels of government lacked either the institutional capacities, the political will, or both, to confront these issues. The relationship between polluters and the federal government in this type of situation is often quite hostile, with lawyers often reaping the inevitable benefits of such communication problems.

Chapter III deals with alternatives to the prevailing view, which are becoming more acceptable in the field. Market incentives and the libertarian view of free-market environmentalism are of current interest. Contracting out with private companies by state and local governments for the purpose of cleaning up pollution is becoming more prevalent, and is used in the base conversion process. Environmental justice is of concern to communities located near polluted military facilities. The intergovernmental ideas of bottom-up implementation and devolution are both important issues in this area and pertain directly to military base redevelopment. To alleviate the scarcity problem for communities affected by base closure, the federal government provides technical expertise and redevelopment grants, and is liable for any environmental damage. This chapter attempts to illustrate that there can be a reconciliation of interests between the public and private sectors, and levels of government, over quality of life issues when the political will to do so is present.

While the first part of the book deals primarily with context and environmental policy theory, the second half takes on the various developmental and remedial issues associated with defense downsizing in the 1990s. Chapters IV, V and VI discuss problems, solutions, processes, and outcomes using comparative case studies and quantitative analyses. A more detailed discussion of methodology is described when necessary. Whereas the first half of the book deals with theory, the second half deals with practice, and the concluding chapter attempts to link the two.

Chapter IV contains a quantitative analysis of survey data that was used to measure environmental variables in the base transition process at 44 of the 115 BRAC cases. I also discuss the most important determining factors and independent variables with regard to policy outcomes. The role of economic incentives as opposed to environmental solutions is discussed and explanations for base conversion failure are dealt with. Environmental variables are demonstrated to have an intervening effect in the process, and can result in significant delays if not effectively addressed. In particular, strategic planning, writing the environmental impact statement, and choosing solutions such as Superfund, parcelization, and BRAC clean-up teams are all necessary preconditions for economic development at former military bases. This argument is further illustrated in the narrative case studies in the following chapter.

Chapter V brings home the notion that environmental degradation is perhaps the most daunting barrier to successful military base redevelopment. Cleaning up contaminated bases is a very expensive proposition, both in the short term and long term. Although the federal government is liable for the cost, many local governments do not have the resources to implement and oversee such activities. The question then becomes how to expand institutional capacities to meet the demands of remediation. Pollution at installations can range from spent shells and bullets, to oil and dangerous chemicals, to hazardous and toxic wastes. Private companies do not want to be held liable for military-related pollution and many people do not want to work or live near contaminated facilities. This raises questions regarding economic development versus environmental justice. Specific types of pollution, the methods used to deal with it, and the associated costs are all discussed. Key intervening variables that delay base transition are addressed in this mostly explanatory chapter. Several case studies that illustrate state, local, and private innovation are provided as examples, including Orange County, California, Austin, Fort Worth, and Lubbock, Texas, Fort Smith, Arkansas, and Jacksonville, Florida.

Chapter VI is the concluding chapter and attempts to reconcile theory with practice, as well as the demands of the market and quality of life. This is where the common threads of the main theory and the two analytical chapters are integrated to form a comprehensive theory of sustainable development policy. Given the willingness of the EPA to operate in a more bottom-up fashion and of the Pentagon to let state and local governments implement clean-up strategies, which they then fund, progress has been made. Though at times the two predominant theories of environmentalism may compete with one another, for the most part they can be reconciled as a logical progression from one set of circumstances to another. Final observations on the process and outcomes, and commentary on future base closure and conversion rounds, are discussed. I conclude that the BRAC process could become a regular occurrence and that it could be further streamlined. Congressional pork and bureaucratic turf wars are not likely to let this happen however.

Note

1. Devolution is, however, part of the context, and may even be a necessary precondition to defense downsizing, as many of the proposed solutions have been experimented with previously to deal with similar problems.

Chapter II

The Environmental Orthodoxy

The literature on environmental policy implementation can be categorized into at least two distinct areas, command and control versus incentive-based sustainability. The first, and predominant theme, which is the subject of this chapter, is command-and-control implementation. In fact, hierarchical, top-down policy implementation had enough widespread acceptance to be considered the operative and intellectual paradigm of environmentalism until recently. This chapter utilizes Chuck Fox and Hugh Miller's (1995) administrative discourse theory to critique the command-and-control paradigm. The structure and behavior of institutions and actors, such as federal agencies, media, environmentalists, and corporations, are most likely to be defined as causal variables in top-down environmental research. Policy outputs (sets of decisions) and outcomes (subsequent impact) are often used as dependent variables.

These actors often behave in accordance with what Fox and Miller characterize as the "representative democratic accountability feedback loop" (Fox and Miller 1995). This characterization is a critique of David Easton's (1953) "black box" model of political behavior and its contemporary variants, such as Paul Sabatier's (1988) advocacy coalitions, among others. The representative democratic accountability feedback loop is the idea that when people have problems or concerns that are salient enough to warrant political action on their part, they will organize and make demands of the political system until something is done to their satisfaction. Rules are made by legislatures, budgets and personnel are allocated, and policies are implemented and evaluated by bureaucracies in a continuous loop, even after the original complaints have long been forgotten.

Easton's (1953) notion of politics is quite mechanistic. Citizens put demands and supports (inputs) into the "black box," which represents political institutions. The system then deals with these demands, makes an assessment regarding resources, and spits out a "policy output" to tackle the problem. These outputs have subsequent impact or policy outcomes, which are then evaluated in the political environment, and return to the black box automatically via the feedback loop as further demands or supports. Generally, if a policy works there will be little to no positive feedback. If there is not much complaining on the part of citizens, then the politicians and administrators operating the black box of government institutions consider the policy to be effective.

W. Douglas Costain and James Lester (1995) discuss three similarly mechanistic models that purport to explain longitudinal (over time) policy change with regard to the environment. They include Schlesinger's (1986) thesis regarding cyclical politics, Sabatier's (1988) policy learning/advocacy coalition theory, and Edwin Amenta and Theda Skocpol's (1989) backlash or zigzag thesis. All three of these models are top-down, rationalist in orientation, and rely heavily on the feedback loop concept.

Briefly, Schlesinger's notion of cyclical politics is based on a generational idea of socialization, where citizens learn their political values and beliefs at a young age, then put them into practice when their generation takes over the reins of government. He argues that this happens in thirty-year cycles, alternating between periods of liberalism (high levels of government action) and conservatism (heavier emphasis on private-sector solutions). With regard to environmental policy, his thesis holds up over most of the twentieth century but falls apart in the 1990s. Despite calls for additional environmental policy in the late 1980s, there was little additional policy to come out of the federal government during the next decade, which should have been characterized by an outpouring of liberalism.

Aside from the Clean Air Act Amendments of 1990, which was passed during the first Bush administration, there was little in the way of major environmental legislation that would have been on par with what was passed in the 1970s and 1980s. Instead, in the mid-1990s Congress was taken over by conservative Republicans, whose leadership has by and large been hostile to environmental policy implementation. The Clinton administration, despite the presence of former environmentalist Senator Al Gore as vice president, largely chose not to focus on such issues except at the very beginning and the very last days of their presidency. Though they created more national parks and negotiated the Kyoto Accords, there was really no ground-breaking environmental legislation passed during their administration. Al Gore's loss of crucial votes in key states such as Florida and New Hampshire to Ralph Nader's Green Party candidacy surely was a result of the Clinton administration's and the Gore campaign's failure to adequately address environmental concerns (Bosso and Guber 2003).

The overt hostility of the George W. Bush administration toward the environment is equally puzzling if one does not consider his close ties to "big oil." Bush's hostile appointments of Gail Norton to head the Interior Department and Christine Todd Whitman to head EPA are reminiscent of the Reagan Administration's appointments of James Watt and Anne Gorush-Burford. Bush's rejection of the Kyoto Accords, which were largely negotiated during the administration of his predecessor, and his revival of drilling for oil in the Arctic National Wildlife Refuge and Florida Gulf Coast were also antienvironmental positions. In fact, George W. Bush is widely seen as being even more anti-environmental than congressional Republicans and Republican governors such as his brother,

Jeb Bush, who oppose such actions. Many Republicans, such as Senators John McCain and (now Independent) Jim Jeffords support greater emissions standards for motor vehicles and oppose the Bush/Cheney agenda, which has clearly stated its preference for energy and jobs over the environment (see Vig and Kraft 2003 for a more in-depth discussion of the second Bush administration's policies). According to Schlesinger's view, politicians from both political parties, such as Bill Clinton and George W. Bush, who grew up during the early days of the modern environmental movement in the 1960s, should have been much more willing to implement protective and regulatory policies than they were or currently are. While Schlesinger's cyclical view may have an intellectually simplistic appeal, it falls apart over time, especially in recent history.

Paul Sabatier's (1988) advocacy coalitions are designed to be as similarly predictable as Schlesinger's thesis. Sabatier's theory is not unlike other pluralist models of behavior in the sense that it is overly rationalistic. It is useful for explaining the "analytically tractable" problems involving competing interest groups. These interests may be purposive (ideological) or, more likely, particularistic (monetary or materialistic). Elites and institutions structure these interests hierarchically, except for those concerned with single-issue policy outcomes (Sabatier 1988). A corollary to the theory is that the more conflict there is between coalitions, the less likely there is going to be change in contemporary policy practices. Under these conditions of conflict, change comes only as a result of exogenous events or actors.

Though Sabatier's theory may include more actors than the traditional "iron triangle" style of subgovernment, which might make it theoretically useful for the study of environmental policy, it is no less top-down or exclusive in practice. Advocacy coalitions are rationalist, fragmenting collections of policy subcultures that are controlled from above by bureaucratic and media elites. Sabatier's theory feeds on the representative democratic accountability feedback loop to maintain its causal orientation. It assumes that purposive and particularistic interests operate a priori or independently of the articulation of public policy problems. This would suggest that these coalitions exist only for the sake of self-interest as opposed to any sense of the public good, or of normative, postmaterial values, such as environmentalism. Or as John Kingdon (1984) might argue, that they concoct their preferred policy solutions prior to the articulation and prioritization of political problems.

The actors involved in Sabatier's policy subsystems involve the same set of policy actors that might take part in a situation-regarding discourse, but are held hostage to the feedback loop. This loop is necessary to his theory as it assumes policy learning through the reiterative process of formulation-implementation-policy analysis-reformulation. In a certain sense, it is cybernetic and static, with little experimentation in the policy learning process. The theory also indicates that subsystems are necessarily comprised of competing coalitions that talk past each other rather than among themselves. This contributes to what Fox and Miller (1995) call the postmodern problematic. If advocacy coalitions do in fact behave in such a manner as to add to the "white noise" of hyperpluralism, then they contribute to the problems of society rather than providing any substantive solutions. Hence, they should be denied their warrants (licenses) for participation.

Due to the focus on competition for particularistic benefits, the units of analysis used in the measurement of this type of behavior differ from those that might be used in the empirical verification of more bottom-up theories. The differences in units of analysis lie between that of the rational actor versus those of the disfranchised (as we would find in situations involving environmental justice), or of more communitarian interests, which weigh the needs of the many versus the wants of the few. The assumptions underlying the behavior of the different units also diverge. While rationality does not preclude altruism, it assumes individual or institutional self-interest. A community exists in the pursuit of some sense of the common good or public interest.

The differences between environmental discourse and the institutional elitism of advocacy coalition theory are clear. Whereas rationality assumes that the behavior of collective actors will in some way manifest the sum of their component parts, Fox and Miller's (1995) discourse theory assumes that a discursive community will take on a dynamic distinct from that of its individual membership. In this sense it is a more "totally organic" experience. In addition, advocacy coalition theory does not address what happens when new and intractable problems appear without the existence of institutional elites a priori to the issue area. By anchoring itself to institutional elitism, advocacy coalition theory precludes itself from the formation of noninstitutional *ad hocracies* (Mintzberg 1994), which come together to deal with unexpected, transitory policy problems.

Besides its institutional entrenchment and self-interest, advocacy coalition theory is time-dependent. That is to say, coalition behavior only manifests itself recursively over time. Sabatier does not trust the existence of coalitions that have not had a decade or more to coagulate. Discourse is more flexible in this regard. One does not need to be bound to a particular institution or set of institutions for a period of time to achieve some long-lasting significance. In fact, such bounded institutionalism is likely to be a barrier to the formation of democratic patterns of administration, rather than an explanation of what happens with regard to local, complex economic development and environmental problems, such as base conversions. Although for obvious reasons I do not ascribe to this theory as a prescription, it may in fact be the case that static institutionalism was the historical reality between the early 1970s and the early 1990s for much environmental policy. Costain and Lester (1995) may value the learning-over-time aspect of advocacy coalition theory, but it should be kept in mind that policy analysis is not limited to this theory alone, nor does it always lead to the most optimal solution. The point of advocacy coalitions may not be to implement policy, but to block it, as is typical of most interest group behavior.

A third theory discussed by Costain and Lester (1995) is the backlash/zigzag thesis of Amenta and Skocpol (1989). This view relies heavily upon a stimulus-response reaction by government to problems in the polity. It is similarly rationalistic to advocacy coalition theory except that it involves an element of class struggle. According to Amenta and Skocpol, policy results from the tension between upper and lower classes, who, in the United States, are represented by one of the two major parties. When one class seems to be getting more than its share, there is a backlash by the other group to make sure its grievances are addressed. This does not seem to apply to environmental policy very well because it does not really account for the crosscutting, bipartisan nature of the environmental movement. Nor does it recognize that the differences between the two major parties are no longer based on class consciousness, if they ever were. It also falls apart in the 1990s for the same reasons that Schlesinger's historical theory does: The backlash against the antienvironmentalism of the Reagan administration did not manifest itself in a deluge of new policies under Clinton-Gore.

Costain and Lester seem to think that of the three longitudinal theories that might explain environmental policy, Sabatier's policy learning/advocacy coalition view does the best job. I would concur, but with the caveat that it only really works as an explanation of government policy from the 1970s to the early 1990s. During this time, policy was debated and implemented by a group of bureaucratic elites and their supporting interest groups in the polity and the media, from the top down. Environmentalism was institutionalized in the EPA and other regulatory agencies because of how the rules were written. Far from learning from their mistakes, however, they seemed to make the same ones over and over with regard to rules and hierarchy, due to the inability to escape the policy feedback loop. This created the environmental orthodoxy.

Fragmented Implementers

Built into the environmental regulatory system is an institutionalized antagonism between the public and private sectors. Environmental interests turn to government to mediate the unintended consequences of private-sector pollution, which then perpetuates itself in a seemingly endless cycle of top-down policy implementation and feedback loops. Economic development is juxtaposed against environmentalism in a "jobs versus trees" version of the zero-sum game. What is left out of this equation is the reality that private interests might be willing to internalize their effluents if given the proper incentives (Anderson and Leal 1991), and that job creation can coexist with a clean environment (Obach 2002). Instead, firms believe they are forced into defending themselves from federal agencies. They do this with armies of lawyers funded with deep corporate pockets (Hird 1994).

But even more critical is the fact that government is just as hostile toward the environment as the private sector (Jacob 1992; Kraft 1996; Smith 2000). This makes for a "do as I say, not as I do" policy of hypocrisy in which the federal government reserves for itself the right to pollute in the name of national security but denies private companies the same right. What is ironic about this is that both factions ostensibly have the same goal—the defense of capitalism. Corporate interests, along with government agencies, are part of the advocacy coalition of organized interests that typically blocks policy innovation. The current policy system—top-down though it may be—is incredibly fragmented, providing numerous venues for blocking reform and for agency capture. This divide and conquer method has served both the corporate and bureaucratic interests very well for many years.

The history of environmental policy implementation in the United States since the late 1960s has been characterized by national-level decision making and enforcement, with state and local governments playing subservient roles (Field 1994; Kraft 1996; Lester 1995; Smith 2000; Vig and Kraft 2003). This has been done for several reasons. The first involves the limitations of U.S. states in the late 1960s and early 1970s to deal with environmental contamination. The generally small jurisdictions of states and local communities are insufficient to deal with macro-level pollution of common resources, such as air and water. Also, they generally did not have a sufficient level of support for environmental protection institutions, if they had such institutions at all (Lester 1995). A third reason is that only the federal government has the power and resources to take on large polluting corporations. Many environmentalists believe that localities are unwilling or structurally unable to deal with contamination (Mazmanian and Sabatier 1983), which is why they choose macro-level venues to pursue policy goals.

Despite the hierarchical characteristics of national environmentalism, policy is not implemented in a smooth, coherent fashion the way one might expect. Instead, policy-making is divided between several federal agencies, which often compete for the same resources and jurisdictions. There are no less than sixteen executive departments involved in this process, including Interior, Commerce, Agriculture, HUD, Justice, State, Energy, and Defense, just to name a few. Also, there are at least a dozen congressional committees that legislate environmental policy for all of these federal agencies (Vig and Kraft 2003, 5-7). Though policy may be implemented from the top down, it is not centralized in the traditionally Weberian (1997) sense. This demonstrates that the conventional conception of centralized, coherent policy implementation at the national level is not the reality.

One could argue that policy aims might be better served if in fact these various agencies and bureaus were centralized, but there is evidence to the contrary. Barry Field (1994) argues that the majority of environmental policies are not well conceived. This means they do not address the problems of polluters or contamination in a manner that is economically feasible. "Not being cost effective, they will end up costing lots of money and having much smaller impacts on environmental quality than they might have had with better design" (1994, 10). So, not only are there cost problems, but there are efficiency problems as well. Further bureaucratic centralization would probably not address these issues very well, because it would be undermined by poor causal theory. The separation of powers in the U.S. system compounds this problem.

Barry Rabe (1994) and Michael Kraft (1995) both make similar arguments. Rabe argues that the reason why programs such as Superfund are so inefficient is because of legislative and bureaucratic incapacity. Fragmentation among congressional committees and subcommittees leads to bad policy (though it may have started out as a good idea) through statutory inconsistencies and a true lack of priorities. Congress may also be guilty of overestimating the ability of EPA to carry out such a difficult set of policy mandates (Rabe 1994). Kraft (1995) adds that the centralizing influences of partisanship and the presidency are not enough to compensate for the trained incapacity of bureaucrats or the continued statutory fragmentation caused by what he terms "subcommittee government." With regard to Congress it would seem that the right hand truly does not know what the left hand is doing.

Walter Rosenbaum (1995) takes on this issue with regard to the executive. In addition to all the cabinet-level departments and independent agencies that carry out environmental policy mandates, there are five presidential staff agencies that do the same. Many of these bureaucracies attempt to avoid agency capture by regulated interests by resisting policy changes and hostile appointments. How do they do this? Rosenbaum argues that they institutionalize environmentalism into their missions, rules, and procedures. Along with this comes institutionalized distrust, exacerbating the already hostile environment between the public and private sectors. Rather than putting into place rules and regulations that foster neutral competency, which was the case for many agencies under progressivism, implementers institutionalize their environmental regulatory preferences.

Congress also engages in the practice of legislative micromanagement to prevent agency capture. The use of a multiplicity of deadlines and statutory requirements, such as "hammer clauses" (mandated reporting) in the 1970s and 1980s, were designed to thwart hostile takeovers of environmental agencies through increased congressional oversight. Also mandated were practices involving citizen participation and granting greater standing to sue on behalf of the environment. EPA is also required to submit 100-150 annual reports illuminating the implementation progress of its programs (Rosenbaum 1995). Who was hammered by these "hammer clauses"? Rosenbaum argues that it was the agencies and the general public, not the polluters as intended. This extreme level of legislative oversight may also be an artifact of many years of Democratic Congresses and Republican executive administrations, which (especially in the 1980s) had different views of environmental regulation. President Ronald Reagan's practice of hostile appointments to regulatory positions probably backfired in the sense that government regulation increased, rather than decreased, during his tenure in office.

A serious problem is the adversarial relationship that the federal government has with polluters, particularly corporations. The disputes between these two sides involve fights over what should be considered public or private interests and what should be regulated or deregulated (McSpadden 1995; Wenner 1993). Although many of these disputes are debated in the myriad of executive agencies and congressional committees, the most important venue in this relationship is the court system. Anywhere from four to five hundred environmental cases are filed each year, with judges playing the role of informal mediators in order to reduce the number of cases actually going to trial (Wenner 1993). The role of the courts as environmental peacemakers has expanded along with the conflict, with judges often playing an important enforcement role. However, litigation and micromanagement by the courts may often compound implementation problems, because courts are venues where people argue about rules, rather than carry out policy. The courts themselves, by design, do not have the institutional capacity to implement policy, as demonstrated by civil rights policy (Mazmanian and Sabatier 1983).

Agencies may also be subject to what is known as backdoor capture. By using the Office of Management and Budget (OMB) to restrict budgets, and to justify expenditures with benefit-cost analysis, the Reagan White House was able to block implementation of regulatory policy. Along with the practice of hostile appointments and redistribution or reduction in personnel, the Reagan administration was able to do a great deal of damage. Between backdoor capture by the White House and micromanagement by Congress, the environmental agencies were somewhat hamstrung in their efforts to get anything done. The implementing bureaucracies are held hostage by too many rules, too much hierarchy, and too much duplication of effort. Not only does this hurt the agencies, but it has effects on the general public as well, in terms of greater health care costs and damage to the overall quality of life.

Since the 1980s, business interests have actively challenged the federal government's regulatory role in the courts at all levels. This counteroffensive is often referred to as the property rights movement, because the legal basis for such litigation is found in the Fifth Amendment to the U.S. Constitution. Their argument is that such over-regulation reduces the value of real (as opposed to personal) property, and hence results in the government's taking away property without due process. This method has been used to challenge legislation which gives environmentalists greater standing to sue, such as the Endangered Species Act and efforts at protecting wilderness and crucial habitats from development (McSpadden 1995).

These problems also exacerbate the seldom-mentioned difficulties associated with federal pollution, particularly when it involves the overly expensive and inefficient Superfund and related programs. Even though the federal government is the supreme environmental regulator, in the early 1990s it was still responsible for contamination at 116 toxic waste sites, 94 of which were military bases (Jacob 1992). There were nineteen such sites in California alone, and seventeen former nuclear weapons manufacturing plants in thirteen states were likewise polluted (see also Smith 2000). What is interesting about these plants is that they are government owned, but operated by private corporate contractors–clearly an indication of the close relationship between private and public polluters (Jacob 1992). Despite the fact that it is liable under its own laws and regulations, the federal government is very slow to take responsibility for its actions. Due to convoluted federal procedures, implementation time and expenses increase, as do chances for mistakes, which are plentiful.

Superfund Failures

Despite the apparent foot-dragging, the federal government publicly accepts responsibility for its pollution errors, and will not permit redevelopment of contaminated sites until they are cleaned up (Steppick 1994). However, federal agencies are slow to engage in environmental restoration activity at their own brownfields. This can become an insurmountable barrier to local redevelopment efforts if proper procedures are not followed (Gottschneider, Hicks, and Donohoe 1994). In addition to the burden imposed on localities, such as having to write environmental impact statements and incorporating delays into strategic plans, they must also be prepared to deal with the lengthy CERCLA (Superfund) process. CERCLA is the Comprehensive Environmental Response Compensation and Liability Act of 1980. CERCLA liability applies to all present and former owners of facilities or property, which is why the Pentagon is liable for pollution at military installations. Of concern to communities is the notion that they could be held liable if property is transferred to them prior to reclamation. They do not want to be caught in an intergovernmental dispute over liability with the federal government because it is likely to be a losing battle. CERCLA, coupled with the Superfund Amendments and Reauthorization Act of 1986 (SARA), provides billions of dollars in funding for contaminated bases, a great deal of which comes out of the U.S. government's general fund and from those polluters who lose federal suits (Smith 2000; Steppick 1994). The causal theory behind CERCLA and SARA is that restoration should be funded by the polluters. But much of the money is instead spent on litigation (Hird 1994).

To the detriment of local communities, Superfund allocations are made on a "worst first" basis. This means that those localities, which have moderate or small amounts of contamination, may be forced to wait while EPA spends its time on the most contaminated cases or litigates with private polluters in an attempt to recoup clean-up costs. According to the GAO, the implementation of the Superfund is in bad need of reconfiguration. There are no less than fifteen steps in the Superfund process between the time a site is identified and the time it is cleaned up (Davis 1993; GAO 1998; Smith 2000). This creates inordinate delays for communities attempting to redevelop local brownfields.

Another federal statute that applies to polluted military bases is the Resource Conservation and Recovery Act (RCRA) of 1976, which involves EPA oversight of hazardous waste disposal and treatment. This program is the forerunner of Superfund, SARA, and the Hazardous and Solid Waste Amendments (RCRA Amendments) of 1984 (Kraft 1996). RCRA mandates that only those companies that can afford the high liability be permitted to dispose of hazardous waste. The RCRA regulation and permit processes are administered by the EPA. Enforcement is financed by a fund utilizing both public (from the General Fund or Superfund) and private money (fines resulting from assessment of liability).

State and local governments are also liable, as are companies who had no knowledge of prior pollution (Smith 2000). This aspect of the law frightens many local reuse authorities and potential tenants of former military installations. However, RCRA also provides recourse for states seeking action against the military through the court system.

These above-mentioned statutes are a problem for several reasons, all of which relate to inefficiency. Having a convoluted process increases the time and expense of implementation. It provides many opportunities for mistakes to be made. Also, it is possible to get bogged down in the process and lose sight of the objective. When mistakes are made, it is difficult to pinpoint when they were made or what the ultimate effects might be. In short, having a complicated, multiple step implementation process substantially increases the margin of error and the time it takes to effectively carry out policy (Cochran et al. 1999; Davis 1993). Additionally, the fact that new industry or state and local governments could be held liable for pollution that they did not commit on a given site is patently unfair. The resulting arguments can potentially result in a lose-lose proposition for all involved.

In what can only be characterized as a scathing indictment of the Superfund process, John Hird (1994) argued that it was an abject failure. Identifying areas that can be reformed are more difficult however. The largest problems according to Hird involved inefficiency, excessive expenditures on transaction costs such as lawyers and consultants, the arbitrary nature of the National Priorities List (NPL)[1] and its underlying "worst first" policy, as well as the lack of clear and consistent remediation standards. Other problems involve costs being passed on to consumers and taxpayers by corporate polluters that have been brought to justice, and through higher taxes. A recent GAO study on the Superfund program adds weight to Hird's argument. According to them, clean-up times of nonfederal sites lengthened to an average of 9.4 years. Much of this is due to the inefficiency of the planning and NPL processes, rather than the actual clean-up efforts (GAO 1998).

All of these problems have contributed to a lack of public trust and confidence, which is often the case when it is perceived that government has taken the people's money and given little in return. To deal specifically with Pentagon pollution, Congress created the Defense Environmental Restoration Program (DERP) in 1986, which was designed to empower the Defense Department to clean up bases without interference from EPA. However, it also operates under a "worst first" policy and has many problems similar to the Superfund. It is also a drag on the base conversion process. "DERP policy fails to facilitate the rapid sale of former military sites that is essential to the success of the base closure program" (Steppick 1994).

So, to summarize, the process of cleaning up former military bases involves a significant number of federal agencies, is intergovernmental, and is procedurally intractable. The question is why does this continue when we know that it does not work? Where is the policy learning that is supposed to take place according to the theory of environmental orthodoxy?

I argue that the problems lie within the representative democratic accountability feedback loop. At some point the process breaks down, regardless of whether we are discussing Superfund, SARA, RCRA, or DERP. One solution might be to consolidate these programs and centralize environmental policy implementation in one agency, presumably EPA. However, that has not seriously been discussed in the literature, and current governmental trends are moving in the opposite direction, except at the state level, where consolidation is being done to counteract agency capture and to combat the budgetary costs of fragmentation. But that is a topic for the next chapter. Whether we have one program or several, the problems with top-down implementation will still be there if they are not addressed appropriately. First, we will deal with the "representative democratic" and "feedback loop" aspects of Fox and Miller's discursive critique, then with the "accountability" and hierarchy issues.

Representative Democratic Feedback Loops

The "representative democratic" and "feedback" aspects of top-down policymaking generally refer to the role played by nongovernmental actors, such as news media, political parties, and interest groups, who often lead and shape public opinion. These groups compete for places near the top of the political agenda with their issues. Many times policies are passed riding a wave of public opinion, but it is difficult to sustain such momentum. John Kingdon (1984) describes this process as the linkage of the policy, problem, and political streams, which then causes a window of opportunity to open. If policy is not legitimized through legislation and budget allocation during this brief period, the policy window will close, and the issue will have to wait again until there is sufficient public support behind it. The feedback loop indicates the public's reaction to such policies or nondecisions.

Fox and Miller (1995) argue that the representative democratic aspects of the feedback loop are mythical, that policy preferences are really dictated from the top down. This would seem to be substantiated by Anthony Down's (1972) rationalistic view of public opinion in environmental policy implementation. Downs argues in his "Issue Attention Cycle" that there are five stages of public opinion with regard to issues such as the environment. They are:

1. Pre-problem. A problem exists, but little attention is given to it by organized interests, the media, or citizens.
2. Alarmed discovery and euphoric enthusiasm. A crisis or some other dramatic events put the problem on the public policy agenda, and demands are made to solve it.
3. Realization of costs. When it is made clear what the policy might cost in terms of additional taxes and regulations, public enthusiasm may dampen.
4. Decline in interest. When the problem is addressed, there is a corre-

sponding reduction of media attention and public scrutiny. Even if there is only a symbolic solution, people may think the problem has been solved.

5. Post problem. The issue is replaced by others on the policy agenda and fades away until the next crisis presents itself (paraphrased from Downs 1972, for a critique see Dunlap 1995).

Downs's view is not unlike the other cyclical views discussed above. Policy elites, riding the twin waves of issue salience (intensity) and valence (consensus), are often happy to oblige voters with new policies. However, these may or may not be effective. They may take the form of symbolic policies or Band-Aid approaches whereby the problem is addressed, but not adequately. Symbolic victories are Pyrrhic in the sense that legislation is passed, but funding and implementation are insufficient. Superfund and its statutory contemporaries might be described in such a way. They are expensive, Pyrrhic policy boondoggles for environmentalists and the public.

Government may also take the opportunity to put off action by issuing a nondecision, or redefining the problem so that a workable solution is more expensive than citizens are willing to pay for. Such is often the case when there is a dispute over development versus the environment. It may be the case that there is really no conflict between business interests and quality of life issues, but if they are defined as an either-or proposition, or zero-sum redistribution, enough people may argue that the costs of regulation are too high. This may postpone action for long enough so that the problem goes into the "decline in interest" and "post problem" phases, where people believe the problem has been adequately addressed, or it just disappears from the public's radar screen until the next crisis.

Accountability versus Discourse

The publication of Mazmanian and Sabatier's *Implementation and Public Policy* (1983) as a response to Lipski's (1976) work regarding new forms of administration resulted in the germination of a whole new debate about the structure of policy implementation at all levels of government. Since it has been twenty years since the original edition was published, perhaps now is a good time to review the applicability of the model in the new era of reinventing government and Pentagon downsizing.

How valuable is the top-down, hierarchical model to addressing environmental degradation and military base redevelopment? This question cannot be answered without taking into account the bottom-up responses presented by Hjern (1983), Elmore (1985), and others. Neither can the context of scarcity of resources in which this debate takes place be ignored.

Some of the most notable features of the Mazmanian and Sabatier model are its hierarchical structure, its causal theory, and its six conditions for successful implementation. The feature of hierarchy in the model is noteworthy only in the sense that it produced an intellectual outcry against it. There are some disadvantages to top-down structures, which will be touched on later, but most bureaucratic models are in fact top-down by nature, and this was particularly true at the time the book was authored. A criticism of the model is that it relies too heavily on statute, downplaying other essential nonstatutory variables, reinforcing the notion of hierarchy. However, current environmental enforcement relies heavily upon statute and its underlying causal theory of liability.

Mazmanian and Sabatier emphasize the notion that implementation is only as successful as the causal theory behind the enabling legislation (statute) in their six conditions. Statute can be a constrictive force on successful implementation, but if done correctly it can also be a boon, allowing regulators additional powers and leeway that they did not have previous to legitimation.

While others choose to focus on the top-down structure and reliance on statute, it makes more sense to examine the six conditions for effective implementation, as that is the real contribution of the work to our theoretical understanding of implementation. To briefly summarize the conditions they are:

1. Problem definition and prescription must be clear and consistent.
2. The causal theory is correct, and target groups can be directly identified and served by public officials.
3. Agencies have the means, both financial and otherwise, to work with the target groups.
4. The administrators are committed to successful implementation and substantive results.
5. There is active support among constituency groups and public officials.
6. Priorities are not weakened by competing problems or policies and the original causal theory is not weakened (paraphrased from Mazmanian and Sabatier 1983, 41).

A common thread that runs through the above conditions is the accuracy of the causal theory, as opposed to emphasis on statute. No matter how well intentioned the legislators are, if they have not properly defined both the problem and the solution, then the power of statute and hierarchy becomes moot. Most environmental enforcement agencies and personnel are committed to working with the public and interest groups to clean up pollution. However, if the enabling legislation does not adequately define pollution, liability concerns, and the desired outcomes, there will be problems with the implementation of clean-up plans, regardless of the implementation structure. What this means is that even top-down theories of implementation expose the weaknesses of current environmental policies, such as Superfund, SARA, RCRA, and DERP. The antagonistic causal theory of liability and "worst first" behind such statutes are what contribute to the high instances of litigation in environmental enforcement, rather than simply addressing issues of cleanup or of internalizing environmental consequences.

A second problem with the model returns us to the notion of hierarchy. Many academics and antibureaucratic, free-market conservatives criticize top-down structures as being undemocratic and unresponsive, advocating more decentralized, bottom-up oriented structures. Bottom-up structures are intuitively pleasing to nonconformists and in situations where hierarchy is clearly not applicable. However, to criticize Mazmanian and Sabatier's model simply because hierarchy is intellectually unappealing is missing a crucial point. The point is not that top-down modeling is a bad thing—it has its uses, the military being the first to come to mind, but by no means is it the only one. The problem with the Mazmanian and Sabatier hierarchical model lies in the inherent limitations of its unidirectional (one-way) causality.

Those in favor of bottom-up organizations have attacked the exclusive nature of unidirectional hierarchies. This is an important aspect of the arguments made by Hjern (1983), Elmore (1985), and Fox and Miller (1995). The flows of information and communication are necessarily multidirectional for a bureaucracy to effectively carry out its mission. This is especially true for base conversion and environmental remediation efforts. The street-level bureaucrats must articulate needs and solutions to the upper-level bureaucrats and legislators, who then provide resources and legitimacy to the relevant agencies. The key difference between the two sides in the debate is not so much the directional flow of communication and authority, as it is the degree of discretion that the street-level bureaucrat should be permitted in participating in the discussion and in making decisions.

Recently the top-down model has come under fire in both corporate and governmental contexts. Large "inflated" bureaucracies, generally modeled in some hierarchical fashion, have become targets of downsizing in a new era of fiscal and economic austerity. Organizations and corporations have had to "work more and cost less," to quote a phrase popular with former President Bill Clinton, to deliver services and to compete with foreign interests. Contrary to previous expectations, there has been no peace dividend—thanks in some part to the costs associated with Pentagon pollution—but policy demands still seem to have increased with the ending of the Cold War.

In order to meet these demands a greater responsiveness is being required of governments. There can no longer be a Wilsonian (1997) division between administration and politics, if there ever was one. Administrators can no longer be isolated from the public by shielding themselves with layers of bureaucracy. Symmetrical pyramids of hierarchical organization are no longer seen as able to effectively implement policy in many instances.

The criticism that top-down structures rely too much on statute for their focus of operation comes chiefly from Hjern's (1983) critique of Mazmanian and Sabatier, who place a great emphasis on law as a guide or blueprint for administration, when it may actually have detrimental effects. "'Clear and consistent objectives in a statute,' assumed to be positively related to the ability to structure and to implementation success, may actually inhibit and hinder the implementing officials from using their leadership skills" (Hjern 1983, 303). Hjern also discusses how Mazmanian and Sabatier downplay non-statutory variables that may affect implementation. He maintains that the two could easily be linked. "A potential for such interrelations is easily detectable in their scheme" (Hjern 1983, 303). The nonstatutory variables discussed by Mazmanian and Sabatier are socioeconomic conditions, public support, attitudes and resources of constituency groups, support from sovereigns, and commitment and leadership skill of implementing officials (Mazmanian and Sabatier 1983, 22). This sounds a great deal like Sabatier's advocacy coalition, which came soon after. But it could also be interpreted to mean something like Heclo's (1977) issue networks or Mintzberg's (1994) ad hocracies, depending on the direction of communication and the transitory nature of the structure.

The second criticism is related to the first, in that if the statutory and nonstatutory linkages are established, then the division between politics and administration is removed. This is true because as seen above, the non-statutory variables represent elements of the political process, while the statutory variables represent the structure of implementation. Clearly, the two interact in reality, but Mazmanian and Sabatier do not indicate a formal relationship in their model, although as Hjern indicates, an informal relationship is implied. Hjern argues that there is a conceptual link between administration and politics, and that link is policy implementation (1983).

Baumgartner and Jones (1993) argue a similar relationship, although they distinguish between image and venue, rather than administration and politics. However, image as they define it is very similar to the previously outlined description of politics and nonstatutory variables. Venue similarly describes the institutional jurisdiction in which the policy will be implemented. Their contribution to the debate is the notion that changes in the way the problem is defined generally will result in changes in the way it is implemented. Hence, policymakers will attempt to utilize venues in which they have the greatest chances for success. Many local conversion efforts involve extensive private-sector involvement, which is why venues and nonstatutory variables become important.[2] The third criticism regarding the limitations placed on street-level bureaucrats is dealt with extensively by Elmore (1985) in his article on backward mapping. In this work he challenges the assumption that policy implementation is controlled from the top down. "The most serious problem with forward mapping is its implicit and unquestioned assumption that policymakers control the organizational, political, and technological processes that affect implementation" (Elmore 1985, 20). Backward mapping alleges that the problem and solution are defined at the lowest level of bureaucracy and are then sent back up through the structure to receive funding and legitimation so that the solution may be effectively carried out. This idea also challenges the premise of hierarchy, claiming that it is not necessary. "The closer one is to the source of the problem, the greater is one's ability to influence it; and the problem solving ability of complex systems depends not on hierarchical control but on maximizing discretion at the point where the problem is the most immediate" (Elmore 1985, 11).

The most obvious implication of this would be the relative freedom and discretion that lower-level bureaucrats would enjoy compared to those under the top-down system. It also places a greater importance on the informal network that facilitates communication in all directions. This could be especially crucial in the formulation of strategic plans for military installation redevelopment. With greater communication and discretion, multiple centers of power are likely to develop, creating polycentric administrative structures, similar to the "technocracy clusters" described by Stillman (1991), or what Fox and Miller (1995) call nascent forms of discourse, which are less hierarchical than Stillman's notion of technocracies.

However, as bureaucracies are downsized and restructured, there still remain several important things to consider. If something like nascent forms of discourse or ad hocracies are to assist in base conversion efforts, who will be held accountable with no clear chain of command? With regard to intractable problems that could pose barriers to conversion, such as environmental contamination, an answer to this question becomes that much more crucial. Somehow a system of accountability must be maintained as protection against lawsuits, duplication and corruption. It seems that with decentralization there is a trade-off between the need for greater communication and responsiveness on the one hand, versus accountability and chain of command on the other. In order to find a more perfect system of administration, these differences must somehow be reconciled. Such an ideal may not be possible. We may just have to deal with the tension between these elements of management.

Elmore seems to indicate that if bureaucracies are structured in a bottom-up fashion, they will in fact be more representative of the will of the people than top-down structures, with positive results for issues such as contamination and base conversion, though this is an inference of mine. The question for Elmore is not one of accountability, but of efficiency and responsiveness. Mazmanian and Sabatier, on the other hand, seem to indicate that it is not the job of bureaucracy to represent the polity, but to implement policy that has been previously legitimized through legislation. In accordance with Wilsonian (1997) assumptions of the separation of politics from administration, questions of justice should be addressed by statute, not by bureaucrats. The fear on the part of top-down advocates in this instance is lack of accountability on the part of street-level bureaucrats who might be given too much discretion and exceed their levels of legitimate power.

How then do we hold polluters accountable for their actions? One way might be through greater openness. Negative publicity can have a great effect on corporate and governmental behavior. Recall the public outcry against the Exxon Corporation (now ExxonMobil after a fit of petrol merger mania) following the *Exxon Valdez* accident in 1989. Such a pattern of un-green behavior can result in customer boycotts, devalued stock prices, as well as lost commodities, market share, and increased insurance liabilities. Public interest groups and concerned citizens, as well as enforcement agencies, can all contribute equally in holding corporate polluters accountable for their actions.

This discursive sunshine policy can also apply to governmental polluters. The environmental justice movement has focused on the location of toxic and hazardous waste sites in inner-city neighborhoods by municipal governments and exposed them in a NIMBY-like manner. This could also happen in communities where the military pollutes. Citizens will complain when military personnel engage in incidents of drunken debauchery and criminal activity, such as rape or assault. To its credit, the military generally responds quickly to such behavior in the interests of being a good neighbor (Sorenson 1998). Rather than hiding behind a veil of national security the military could be more open about activities that might affect the health and well-being of neighboring communities. For this to work, there has to be a greater reliance on ethics in government, and on societal (or perhaps peer) pressure. The antagonistic relationship between government and the private sector must also be addressed. There can be an agonistic tension without open hostility.

It should be apparent that federal top-down policy management is very complicated and problematic. There are too many agencies and committees with competing interests and limited turf. Though it has done some good, environmental policy is neither cost-effective nor efficient. The courts have become burdened with mediating and enforcing policies between adversarial agencies and polluters. Also, little has been done to address the issue of what happens when the government itself becomes an environmental outlaw, as it does at military bases and Department of Energy sites. Greater openness, such as might be found with an environmental discourse, might be an answer. Perhaps also greater market incentives could be found to encourage corporations and government to internalize the unintended consequences of their actions before punitive measures have to be taken. Last, but not least, polluted facilities could be cleaned up and reused, creating greater sustainability. These issues will be addressed in the next chapter.

Notes

1. The National Priorities List is part of the CERCLA process in which states assist the EPA in identifying and arranging the most contaminated sites into a triage-like hierarchy. Those locales that are the most polluted are placed at or near the top of the list for remediation.

2. In a related piece, which appeared in the journal *Administration & Society*, Dan Krejci and I take the Baumgartner and Jones argument a step further and illustrate the formation of entrepreneurial venues in the base retention process (Hansen and Krejci 2000).

Chapter III

Policy Implementation Alternatives

Are there alternatives to federally controlled, top-down environmental policy implementation? The answer is yes, there are at least three: devolution, free-market incentives, and civic environmentalism. Elements of all three types of policies can be utilized at former military bases, and often are used in concert. Regardless of policy preferences, if any of these types of solutions are to realize implementation success, an intergovernmental environmental discourse must take place between all concerned institutions, interests, and individuals. This is needed because each situation is unique, requiring individually tailored solutions to deal with the myriad contamination problems at hand. It is also necessary to coordinate the large number and different levels of federal, state, local, and private actors involved in the redevelopment and clean-up processes.

The previous chapter discussed the difficulties of having a simplistic, one-size-fits-all, administrative orthodox solution to environmental problems. However, it is much easier to criticize than it is to provide better ideas. The implicit argument in the aforementioned critique is that there needs to be a contingency theory capable of addressing the complexities of different cases and situations. While the orthodox view might be good at assessing contamination problems and liability, subsequent policy fragmentation derails effective solutions. The difficulty lies in effectively finding and implementing adequate options that match the uniqueness of degradation at specific locations. This is especially true when the federal government is the polluter, which is the case with military-related contamination. When this happens, the old tenets of the environmental orthodoxy must be replaced with new assumptions.

In her ground-breaking work *Systematic Thinking for Social Action*, Alice Rivlin (1971) advances such a contingency theory, which I apply to the analysis of greening military brownfields. Rivlin argues that with regard to social policies, primarily education and health care, the U.S. government has not encouraged enough of the policy experimentation that is inherent in the federal structure. Likewise, it has not encouraged enough incentives to enhance programmatic effectiveness or accountability. Some thirty years after the fact, many elected officials, analysts, interest groups, and others continue to make similar arguments regarding a wide range of policy alternatives and incentives designed to strengthen governance.

Experimentation and Accountability

Rivlin (1971) presents three models to potentially improve social service delivery: decentralization, market incentives, and community control. While each one might not be a panacea, when taken together they can provide potent policy solutions. Key to making these models useful are two conditions—performance measurement and continuous experimentation (1971). These two conditions are precursors to the New Public Management views of enforcing accountability through output assessments and continuous improvement (for critiques, see Kettl 1994; Behn 2002, for supporting arguments, see Osborne and Gaebler 1992; NPR 1993). What Rivlin's models and conditions take into account are the complexities of policy implementation and the difficulties of matching problems with adequate solutions.

Decentralization is the breaking up of consolidated administrative units into smaller, more localized or regional units. The federal government is good at receiving and distributing money, but less efficient regarding complex administrative procedures (Rivlin 1971). Since taxpayers are reluctant to believe that the federal government uses their tax dollars effectively and efficiently, officials local to the problems must be given the power to decide. What she does not discuss in much detail is the role of state governments in service delivery, except to say that they are subject to capture by power elites. To the extent that this happens (and it does), it must be countered through strict accountability (1971).

Rivlin's market model is an extreme, if currently popular, form of decentralization. Many libertarians and fiscal conservatives argue that social service delivery would be more effective and efficient if administered through private companies attempting to increase profits by better serving clients (Rivlin 1971). This is hardly a new idea, however. Government reformers from the Progressives (Wilson 1997; Taylor 1997), to the contemporary reinventors and privatizers (Osborne and Gaebler 1992) have argued that government would work better if it was structured, and functioned, more like the private sector.

The difficulty of hiring private agents to perform public functions arises from a difference in goals, as demonstrated by Ronald Moe's (1997) discussion of the *McCulloch v. Maryland* (1819) ruling. While government agencies perform functions deemed to be in the public interest, private corporations exist to generate profit for their owners. These ends would seem to be at odds with each other when governments contract with private institutions for service provision. Moe argues that the *McCulloch* ruling demonstrated the limits, "both constitutional and political, to privatization ventures" (Moe 1997, 460).

While there are differences in goals and legal limits to privatization, performance can be assessed by measuring outputs, and by ensuring proper procedures are followed. Such benchmarks, as well as public-sector ethics, must be written into the contracts with private entities (see Kettl 1994 for a discussion on the discontinuities involved with principal-agent relationships). However, the debate over the benefits and pitfalls of privatization often clouds another, perhaps more useful, view, that market incentives can provide the proper impetus for better performance, without compromising public-sector goals and ethics, the purpose of market incentives being "to identify and reward more effective management" (Rivlin 1971, 350).

A third model Rivlin discusses is community control. This is a more intermediate form of decentralization designed to include those, usually poor and minority groups, who are typically excluded from political enfranchisement (Rivlin 1971). Such communities are also likely to be the clients or consumers of social service delivery organizations. In a system where community control is utilized, service providers must be held accountable, not only to the taxpayers and the federal government, but also to the community they serve (Rivlin 1971).

While Rivlin does not deal with environmental policy, her three models have been utilized experimentally to help address pollution concerns at different times and places, as seen in the literature. Decentralization has seen states gain greater power in administering environmental regulation through the process of devolution. Market-based incentives and privatization have been used both to prevent pollution and to clean it up. Local control has been advocated in situations involving environmental justice, where minority communities have been unfairly singled out as repositories for waste, because they are poorly connected politically.

Rivlin advocates that these three models be used in conjunction with one another for effective service delivery, as each one by itself had certain drawbacks. As a corollary, I argue that the same is true for environmental policies. When government fails to follow its own policies, it must be held accountable through greater levels of democratization, and participation. By allowing citizens, lower level institutions and private interests into the process, problems can be identified and dealt with in a more open fashion. But the method by which this happens is perhaps even more complex than Rivlin, or many others, realize.

In the intervening decades since the advent of macro-level environmentalism, state and local governments have built better institutions for dealing with contamination. Some of these, California for example, maintain top-down regulatory policies that may not be all that different from national practices (Mazmanian and Sabatier 1983), and may perhaps be even more extensive. Others have turned to more innovative solutions in recent years (Lester 1995). Many states began with creating a state-level mini-EPA, often by consolidating existing agencies to reduce or prevent capture by special interests, and then branched out in other ways to increase their sphere of influence (Ringquist 1993).

Incentive-based sustainable development is designed as a way to harness market forces in an attempt to mediate environmental problems. Sometimes called free-market environmentalism (Anderson and Leal 1991), this view is very similar to the critique provided of the environmental orthodoxy in chapter II. Local power is also utilized to protect minority communities and those directly affected by their proximity to pollution sites. The environmental justice and civic environmentalism movements are examples of these (Shutkin 2000; Ringquist 2003).

What the views of state, market incentives, and local control all have in common are some level of policy experimentation, adaptive (rather than orthodox) responses to environmental crises, decentralized implementation systems, and a more centrist or moderate ideological view overall. Essential to making alternative views effective is coordination among the different players through open communication, or to put it another way, administrative discourse.

In practice, combinations of state involvement, contracting out, and local control are utilized to address the problems of environmental externalities at former military bases. The extent to which each is used depends on state policy, the scale and nature of the pollution problems, and the degree of sophistication present in the local base conversion authority. At issue is the ultimate pace of redevelopment. If military-related contamination is not addressed and rectified near the time of the closure decision, then the local economy stands to suffer from unemployment and lost revenue. This predicament has the unintended effect of turning local business interests and governments into environmentalists, because if they fail to act, the repercussions are potentially devastating for their communities.

Devolution

Devolution involves the distribution of power from the federal government to the states and, often, local governments or special districts. In the United States, devolution as a practice has only been partially successful, as the federal government has been unwilling to provide sufficient resources to the states in order to effectively carry out policy (Hedge 1998; for a comparison with Britain, see Rohr 2002). Hence, states have had to find their own resources in order to meet federal mandates. State environmental policies originated as a way to assist federal implementers. But since the establishment of federal devolution beginning with the Reagan administration's New Federalism, states have become more autonomous and innovative in the environmental realm (Hedge and Scicchitano 1992; Davis 1993; Ringquist 1993; Hedge 1998).

Central to understanding devolution is the resurgence of intergovernmental relations, both in theory and in practice. While thirty-five years ago it was thought that state governments had lost their relevance by being blended into the greater system (see Grodzins 1997), beginning in the 1970s, they once again became important to meeting the policy demands of citizens. Deil Wright's (1988) work on the subject prompted a renewed interest in the study of how states interact with other levels of government in the United States.

State and Local Activism

Though states are perhaps appropriate venues to house environmental policy implementation because of their size and level of sovereignty, policy can be further decentralized. Despite very limited jurisdictions, local-level enforcement of environmental protection has become more widespread in recent years, often with state encouragement and legitimization (Urban 1992; Somma 1994 and 1998).

State and subsequent local environmentalism may provide solutions for federal pollution where the federal government refuses to take action. Even states that have traditionally poor environmental records such as Texas (Ringquist 1993; Lester 1995) have expanded their efforts in recent years. Texas provides us with a good example because it is somewhat typical of environmental agency consolidation in other states, and because it is home to several of the base conversion cases discussed in chapter V.

Part of the reason for increased environmentalism in Texas is the decline of the energy industry's political importance in the late 1980s (Hill and Mladenka 1997). Much like the federal government prior to NEPA, the state of Texas maintained a patchwork of agencies, many of which were weak, captured by the oil industry, and under funded. However, the recent trend among states over the past two decades is to consolidate such agencies and give them the means necessary to carry out effective regulation with a minimum of bureaucratic inefficiency. The election of Ann Richards as governor in 1990 provided the catalyst necessary for this to happen in Texas (Ringquist 1993).

By 1992, several important solid and hazardous waste programs transferred from other agencies to the Texas Water Commission (TWC). Also, the TWC's power of enforcement over municipalities was expanded (TWC 1992). The following year the Texas Air Control Board (TACB) was given increased powers under the Texas Clean Air Act. Immediately thereafter, the TWC and TACB were merged into the new Texas Natural Resource Conservation Commission, or TNRCC, which became the umbrella organization for environmental policy implementation in the state (TACB 1992). Though its detractors referred to TNRCC as a train wreck it was cost-effective and reduced duplication of effort and agency capture. The blueprint for this consolidation was taken from that of other states in an attempt to bring Texas up to date with the rest of the country (Hansen 1995).

Institutional Capacity and Political Will

Evan Ringquist (1993) argues that state-level consolidation of agencies and adequate funding are necessary to maintain effective regulatory enforcement. According to him, the biggest problem states had with regard to environmental protection in the 1970s and 1980s was the creation and maintenance of a large enough institutional capacity to get the job done effectively. With his integrated theory of state policy, Ringquist measures states' institutional capacities as well as policy outcomes to determine a causal relationship between state-level policy-making and reductions in air and water pollution. He concludes that limited institutional capacity restricts a government's ability to handle policy problems (1993).

I agree with Ringquist's assessment regarding capacity, although as Lester (1995) argues, political will is also important (see below). But Ringquist does make some attempt to assess political will by including public opinion as a variable in his model. I also add that those state and local governments that have limited capacities are by necessity going to attempt to find innovative ways to increase their institutional capacities, either on a temporary or permanent basis, by contracting out or by engaging in policy experimentation, when confronted with unexpected contamination. The potential developmental costs of ignoring the problem are too great.

While much of the argument in favor of devolution to states and localities is ideologically based, there are practical reasons for doing this as well. For one, state and local governments have become disillusioned with federal control (Vig and Kraft 2003) and prefer to pursue solutions that better fit their specific situations. Barry Rabe's (2003) notion of Environmental Federalism indicates that in recent decades states have substantially increased their regulatory efforts and capacities and that they are willing to oversee many areas of environmental concern without federal compulsion. States have made progress with regard to pollution prevention, regulatory integration, economic incentives, emissions trading, and mandating pollution risk disclosures. There still remains a great deal of variance among the states with regard to capacity and commitment, however (Rabe 2003).

This variance between states is probably the biggest downside to devolution, and contributes to the potential for agency capture by regulated interests. Lester (1995) categorizes states in something of a matrix according to the level of commitment (political will) and financial resources (institutional capacity) directed toward the implementation of federal environmental protection policies. They are the Progressives, the Strugglers, the Delayers, and—the most hopeless—the Regressives. There continue to be important regional differences associated with these categories, which he does not really delve into much in his discussion, but they are relevant nonetheless.

Most of the states in Lester's Progressive category are the ones we would expect—those on the West Coast, including California, Oregon, and Washington, and those in the Northeast, such as Maryland, Massachusetts, Michigan, New Jersey, and Wisconsin (Lester 1995, 54). Interestingly enough, one southern state—Florida—makes his Progressive list. Undoubtedly this is because so much of its tourist economy depends on maintaining a clean environment. Florida's Department of Environmental Protection (FDEP) has substantial enforcement powers compared to neighboring states. This state is also worthy of note because it is home to one of the cases discussed in chapter V.

The Strugglers include those states with the necessary political will, but inadequate resources to deal with problems sufficiently. Not surprisingly, these are states with smaller populations, and hence smaller tax bases, such as the Rocky Mountain states of Colorado, Montana, and Nevada, as well as the smaller states in the Northeast and Midwest (Lester 1995, 54). The southern state of North Carolina was placed in this category as well, apparently because it ranks eighteenth in the Green Policy Initiative index, which provides some indication of political will (1995, 54).

Lester's third category, the Delayers, includes those states with an adequate institutional capacity, but an inadequate political will to carry out federal mandates or to implement any of their own. Not surprisingly, the Delayers were those states where industrial or energy interests have traditionally reigned supreme over the state governments, such as Alaska, Texas, Oklahoma, Louisiana, Ohio, Pennsylvania, and West Virginia, among others (Lester 1995, 55). While these states have had environmental protection bureaucracies, they have been unable or unwilling to respond to pollution-related problems. While he does not make the claim, it seems apparent from his data that agency capture must be a factor in several, if not all, of these states.

The final category of Lester's state organizational matrix is the Regressives. As one would expect, they have neither the financial resources nor the political will to carry out environmental protection policies. It is in such places that employment versus the environment has been the most contentious of debates. Regressive states included those in the desert Southwest and the Deep South, as well as farm states such as Nebraska and South Dakota (Lester 1995, 55). Many of these states are among the most poor and rural in the country, such as Kentucky and Mississippi, and will continue to attract polluting industries because greener companies have little desire to move there (Lester 1995).

But why is this important to military base redevelopment? The reason why there is variation among states in the devolution of environmental policy, state capacity, and political will is because those states that do not place much priority on contamination issues are likely to contribute to developmental delays. Or to put it another way, because a state does not make environmental quality a priority, it may not fully recognize the potential pitfalls in the causal chain of redevelopment that may result from ignoring military-related contamination, such as liability, health problems, lost revenue, and long-term unemployment.

Market Incentives

Free-market environmentalism pays particular attention to the processes that generate solutions to environmental problems. Anderson and Leal (1991) assert that market incentives are more effective than political means to these ends. "By linking wealth to good stewardship through private ownership, the market process generates many individual experiments; and those that are successful will be copied. The question is not whether the right solution has been achieved but whether the relevant trade-offs are being considered in the process" (1991, 5).

Cost Reductions

The preceding passage is particularly thought provoking because it implies one of the problems with implementing free-market environmentalism—transaction costs. Transaction costs are essentially the costs of doing business and have many sources. In addition to production costs, they include tariffs, regulations, contract negotiation and enforcement, risk, and free rider problems. Cross-national agreements in recent years have attempted to reduce transaction costs. Some examples are the European Community's switch to a common currency—the euro—and the North American Free Trade Agreement (see Anderson 1993 for an in-depth discussion of NAFTA and the environment).

Another significant barrier to effective implementation of free-market environmentalism involves externalities, or unintended consequences. How can the market control the externalities of capitalism? These are the very problems about which environmentalists complain—that capitalism is the reason for pollution. As a solution, environmentalists often promote governmental regulations, or command-and-control measures, as Anderson and Leal call them.

Ideally, externalities should be internalized. Just how this is to be done in practice largely remains to be seen. However, reducing the antagonism between government and business so that they may work together is a necessary precursor (Anderson and Leal 1991; Mitchell and Simmons 1998; Vig and Kraft 2003). Another is creating demand for markets that conserve natural resources, or what I like to call green markets. Rogene Buchholz, Alfred Marcus, and James Post (1992) demonstrate how private or not-for-profit entities have expanded into the area of rain forest preservation, oil spill prevention, and reclamation in pursuit of greener, more responsible markets.

Myrick Freeman (1998) recommends environmental taxes, fees, and the trading of pollution allowances, as other interesting possibilities. Yet another option is to create public-private partnerships, where government agencies engage in reclamation, while polluters reimburse the costs. This was the idea behind the creation of the Superfund legislation, but thus far it has not shown much success in recouping expenses because of disputes over liability assessment (Hird 1994). But if the process could be streamlined, perhaps it could still be of use in enforcing government agency compliance with the law.

Privatization

Another solution is privatization. Privatization is sometimes incorrectly assumed (by both proponents and detractors) to be the replacement of government service producers and providers with private-sector service producers and providers. However, while this is sometimes the case, often it is not. Most privatization involves contracting out—but that can take place in the public, non-profit, or private sectors. More often than not, privatization actually involves contracting with another government entity, but can also involve cooperative endeavors that cross the boundary between public and private sectors of the economy (Ferris and Graddy 1986; Morgan and England 1988).

In environmental policy, there are those who think the solution to the problem of overdeveloping the commons would be to privatize all such resources. William Mitchell and Randy Simmons (1994) hold this view and argue that the reason why governmental regulation of the environment has been inadequate is because private property rights have not been appropriately defined or distributed. Pollution markets, on the other hand, have the potential to deal with the problem as opposed to just the symptoms, which is typically the focus of reactive governmental regulatory policy. "Markets for pollution provide a way of requiring polluters to pay the costs they impose on others" (Mitchell and Simmons 1994, 233). They also argue that this same logic can be applied to water markets and wild animal conservation (1994, 234).

John Dryzek and David Schlosberg (1998) attack the free-market views of Terry Anderson and Donald Leal (1991), as well as Mitchell and Simmons (1994), as extreme hard-line market liberalism. Instead, they are more sympathetic to the quasi-market view of Gooden (1998), which holds that incentives, or "indulgences" as he characterizes them, would be most effective if managed by government so as to maintain an even playing field. Gooden maintains that it is difficult to privatize fluid common resources; therefore, the government must assume the role of owner, for lack of a better alternative. The notion that air, water, forests, and the like cannot be owned by individuals is not unlike the belief of many American Indians, who found the English notion of holding land as private property virtually incomprehensible. Instead, Gooden suggests such things as green taxes as incentives for polluters to internalize their effluents (1998).

Clark Gibson, Fabricel Lehoucq, and John Williams (2002) discuss early efforts at assigning private property rights to manage forests in Guatemala and find that market liberalism does not deliver as promised. They find that public or common institutions and law enforcement are likely to have more beneficial effects than private management. They argue that community ownership of common resources can be just as effective, and sometimes even more so, than privatization.

However, an important point, which is only alluded to in their article, is that Guatemala, along with other Latin American countries, has a tradition of holding land in common—the *ejidos*—as part of their culture. This practice predates colonial times and is part of the regional American Indian society. Perhaps political culture plays more of a role in determining policy preferences than has been addressed in the literature. If this is the case, there will be substantial localized effects, not only when it comes to conservation of natural resources, but also when different solutions are proposed to deal with contaminated property at former military installations in the United States.

Carman (2002) describes how interest groups can take advantage of pollution allowances as another type of political strategy. But unlike lobbying or litigation, buying allowances creates an avenue for direct action. By analyzing environmental interest group behavior in the sulfur dioxide market created under Title IV of the 1990 Clean Air Act Amendments, he concludes that such behavior is rational for those involved, because they realize high returns on their investments (Carman 2002). The implications of his findings are that pollution allowances are likely to be expanded to other areas of concern besides sulfur dioxide.

Brian Cook (2002) conducts an analysis that indicates somewhat different findings from Carman's. Cook argues that while business interests support market-based air pollution control, as we would expect, state regulators surprisingly are in accord with them. Environmental groups tend not to support such regulatory reforms (Cook 2002). However, Carman asserts that the older interest groups are likely to support traditional command-and-control policies, while newer groups were more likely to try something innovative. Carman's view does not necessarily contradict Cook's, but clarifies the notion that there is an expanding diversity of environmental groups seeking different policy preferences.

The implications of market incentives for military base conversion are such that there may be a wealth of possible policy solutions to the conundrum of military pollution besides traditional methods. For localities or even small states, developmental coalitions with business interests and environmental groups are distinct possibilities. This is important because these factions are generally seen as being in opposition to one another, rather than potential business or political coalition partners. Privatization or contracting out for cleanup is a distinct possibility in this context as a way to increase institutional capacities.

Civic Environmentalism

Civic environmentalism combines elements of community development with environmental justice (Shutkin 2000). While environmental contamination is sometimes an issue with regard to military base conversions, community development is essential in almost all cases (Hansen et al. 1997; Sorenson 1998). There is substantially more literature on environmental justice than civic environmentalism because it has been discussed for a much longer period of time—since the 1970s—and because it comprises not only developmental and degradation issues, but also civil rights. The extent to which environmental justice issues are addressed in the community-building and economic development processes is what concerns this analysis.

Assessing Environmental Justice

I think that community development and control is implicit in the literature on environmental justice, because by definition it involves localized groups, minority or otherwise. Hence, there is probably nothing too radical about civic environmentalism. Besides being the latest incarnation of the justice movement, it addresses community power more explicitly, which is useful for our present discussion. Much of the extant literature on military base redevelopment argues that local control is an essential element of the reuse process, regardless of the solutions chosen to deal with environmental externalities (Dardia et al. 1996; Hansen et al. 1997; Sorenson 1998).

Ringquist (2003) evaluates the environmental justice movement and concludes that the evidence regarding the systematic effort to pollute minority communities is inconclusive. However, there is substantial evidence to show that pollution risks are higher for racial minorities. Manuel Pastor, James Sadd, and Rachel Morello-Frosch (2002) demonstrate that Latino children in Los Angeles are more likely to attend schools near hazardous facilities. Jeremy Mennis (2002) shows that hazardous locations are linked to lower socioeconomic status in Pennsylvania. John Pine, Brian Marx, and Aruna Lakshmanan (2002) provide evidence that African Americans are adversely effected by proximity to pollution sites in Lousiana's infamous "cancer alley," which is an eighty-five mile stretch of land on either side of the Mississippi River between Baton Rouge and New Orleans. Potentially, the list goes on.

Regardless of the evidence, Ringquist asserts that there are conflicting goals in the movement, leaving observers to wonder whether it is truly focused on environmental quality, or if social justice, equity, access to health care, and the like are not the real issues. He concludes by saying that to adequately address environmental justice, key terms and concepts such as discrimination and equity in the environmental policy context must be better defined and measured (Ringquist 2003).

Geographic Information Systems (GIS) is a tool that attempts to get at some of the measurement problems Ringquist mentions. It is being used more frequently with interesting results for justice studies, especially when combined with more conventional statistical methods (see Pastor et al. 2002; Mennis 2002; Taquino et al. 2002; Pine et al. 2002). Though to my knowledge GIS has yet to be used in academic studies of military base readjustment, it is certainly applicable to such situations, especially where environmental issues are pertinent. This is an interesting topic for future research.

Robert Bullard (1998) argues that environmental injustices are caused by racism first and pollution second. He argues that even the mainstream environmental movement has engaged in its own brand of racism, due to its composition (mostly white, upper, and middle class) and its focus on lobbying institutions, as opposed to direct action. This is what Jack L. Walker (1991) referred to as "inside" versus "outside" interest group strategies. While Bullard's angered view is somewhat typical of much of the writing on the topic, he also points out that minority communities are extremely vulnerable to "job blackmail," which is the notion that lower-class communities must decide between jobs and the environment, with the environment typically losing out. This is especially the case in those Regressive states that Lester (1995) talks about in his chapter on federalism. Bullard discusses local grass-roots efforts to block the location of polluting facilities in several cities, but does not go into a great deal of depth regarding the politics of community development.

Local Control

By the same token, the community development literature does not really deal with the issue of environmental justice, but does explain how in many instances minority communities are fostered and included by local governments. However, this is a relatively new phenomenon. The "urban regime theory," or local consociational model, may be the most democratic view of community development. Clarence Stone and Heywood Sanders (1987) argue that many cities are governed by a coalition of business and racial minority groups. This is mostly a phenomenon of the post-civil rights era, when previously disadvantaged groups were enfranchised and allowed to have a voice in local affairs.

Stone and Sanders accept that there often are elite-dominated, corporate-entrepreneurial cities, such as those described by Paul Peterson (1981) and Nelson Polsby (1980). They argue that stratified systems may be a necessary precondition to the multiracial urban regime because of the practice of at-large elections. These were designed as part of the Progressive era reforms to stop political party corruption, enabling business interests to take over local affairs, because they were the only ones who could typically afford to run citywide campaigns. The civil rights movement resulted in the restoration of electoral districts or wards in many cities, enabling racial minorities to join the governing coalition through neighborhood bases of support. Many cities have a combination of at-large and district elections to ensure that everyone is represented in developmental decision making (Stone and Sanders 1987; Herson and Bolland 1998). This would largely not have been possible, however, without federal involvement in implementing the *Brown* rulings of 1954 and 1955, or the 1964 Civil Rights Act and 1965 Voting Rights Act (Mazmanian and Sabatier 1983).

Taking the urban regime theory a bit further, Michael Pagano and Ann Bowman (1995, 37-38, 43) argue that there are specific things with which the governing regime concerns itself. For them, normative decisions are just as important as substantive ones. Symbolic politics play an important role in the urban regime. They argue that every city has a normative self-view, and also a view that it would like to portray to the rest of the world. This involves how it would like to develop, as well as how it would like to be perceived. These ideals are what they call "perceptual orbits." These orbits function like conceptual lenses in the decision-making process, eventually leading to tangible developmental policies (Pagano and Bowman 1995, 111).

There is also an urban regime view of these tangible policy outcomes, which Pagano and Bowman refer to as the "tax-services equilibrium" (1995, 139-141). The city's normative view influences and is influenced by the constellation of issues concerning revenues and expenditures. Typically, the normative developmental decisions are designed to enhance revenue, which then translates into additional expenditures for citizen preferences. The regime also ensures that the views of both the business and minority communities are addressed in this process, although in many places this arrangement did not bear fruit until the 1990s, when inner-city gentrification movements resulted in the revitalization of minority-dominated downtown areas (1995, 110).

What can happen according to Stone and Sanders (1987) is that the tangible policy outcomes can be distributed to the business faction, while the minority community is given symbolic victories. In such cases it is necessary for the minority faction, which is often the junior partner in the coalition, to press for financial benefits, in addition to symbolic victories. If done knowledgeably and strategically, the practices of combining normative and tangible financial policies can provide for the redevelopment of contaminated sites in inner cities, dealing with both environmental justice and sustainable economic development problems simultaneously (Fulton and Shigley 2000).

Synthesizing Environmentalism with Community

The emerging notion of civic environmentalism combines the environmental justice problematic with community development and control. According to Shutkin (2000), by combining concern for the environment with the establishment of local networks, a larger constellation of issues can be addressed and perpetuated. Unlike those environmental justice advocates with a left-wing bent, Shutkin is not opposed to economic development, provided that it is sustainable, socially responsible, and green. Shutkin's model provides six core concepts of civic environmentalism that resemble in part the practices utilized in military base conversions. They include democratic participation, community-based planning, environmental education, industrial ecology, environmental justice, and the notion of place (Shutkin 2000). The civic environmental model works well in describing the military base conversion process.

The first of Shutkin's core concepts is democratic participation in decision making. This is necessary to break the elite domination or capture of local institutions. Quite frequently local governments do not have the requisite knowledge or expertise to deal with complex developmental or environmental problems. When crises to a locality's way of life become apparent, what is often in order is a reassessment of a community's perceptual orbit and tax-services equilibrium. Grass-roots involvement of the normally passive citizenry may demand a seat at the table where the big decisions are legitimized. This process is known as inclusion in the base conversion vernacular, and is encouraged by the federal government in those cases where local elite systems are reluctant to increase the level of participation of the general public. Localities that accede to federal demands regarding democratic participation are rewarded with developmental grants (Morgan 1995).

Shutkin's second core concept is community and regional planning. This concept explicitly deals with the determination of what kind of future a community wants for itself in the mid to long term. Or, in the community development vernacular, a local perceptual orbit is negotiated, along with the means to attain it. Strategic planning, which is discussed in great detail in chapter IV, is essential to dealing with the complexities of base conversion. Suffice it to say, without a credible strategic plan, efforts at local democracy are suspect, and outcomes are delayed if not altogether sabotaged. The federal government rewards completed strategic plans with additional grant money, and often the conveyance of the former base property to the community.

The third core concept of civic environmentalism is environmental education. While Shutkin focuses mostly on the efforts to educate children and citizens regarding the danger of pollution, such practices can also inform developmental discussions. When military-related contamination emerges as a barrier to strategic planning and redevelopment, citizens and decision makers must be made aware of liability, potential dangers, and most importantly, solutions that expedite the clean-up process.

Civic environmentalism's fourth core concept is industrial ecology. This is more commonly referred to as "clean production." This is a good deal like green markets, which demand conservation along with the provision of jobs and tax revenues, but without the environmental degradation of traditional smokestack enterprises. The high-tech sector is an example of this. Many former bases are now home to companies that are engaged in light manufacturing and assembly, education, and the like, which are kinder to the environment than heavy industry. Evidence for these claims is provided in chapters IV and V.

Shutkin's fifth concept is environmental justice, or, more accurately, social equity. This of course is much more difficult to achieve in a society in which inequality of wealth is the norm rather than the exception. However, in practical terms, if it means that no one sector of the community is singled out to house a polluting industry, then perhaps equity of a certain kind has been achieved. With regard to environmental justice and military bases, the federal government has not systematically located contaminated facilities near poor or minority communities. Quite the contrary, most of the domestic base infrastructure was constructed during World War II in sparsely populated rural areas, and only later did the cities grow around them. While there have been instances where military contamination went off base and polluted the surrounding area, this is an unintentional occurrence. For the most part, the military wants to be a good neighbor to its local communities (Sorenson 1998).

The final core concept of civic environmentalism is the importance of place. What this means is an affection toward a particular locality. By extension, Shutkin would probably agree with the assertion that the stronger people feel about their place of residence, the more likely they are to take care of it. This is similar to the notion of proprietorship that goes with privatization, except Shutkin does not require ownership for people to have the political will or the capability to value local environs. Instead, he views local resources as something that all citizens hold in common, in attitude, if not in fact (Shutkin 2000). Again, this is not really that much different from the urban development concept of perceptual orbits. Citizens will take into account the quality of life when making developmental decisions. This means evaluating the possible consequences to the environment, along with the potential economic benefits to the private sector and to local governments.

Environmental Discourse

Rivlin's (1971) view of service delivery calls for a determination of solutions based on an assessment of what is the most appropriate for a given situation. Her contingency view of policy involves finding the appropriate level of decentralization, either state, local, or market, or potentially combinations of these. What is left out is a process for how to determine the most appropriate solution locus. Presumably this involves some level of democratic decision making, but this is not explicit. What should be apparent from the running commentary in this book is that administrative discourse is an effective and appropriate method to address such decisions. Two good examples from the environmental policy literature illustrate how this can be done in practice. Douglas Torgerson (1990) discusses how discourse can alleviate problems associated with policy fragmentation, while Kai Lee (1989) provides a case study of the Columbia River Basin where discourse was used to mediate turf battles between states, localities, public authorities, private interests, and Indian tribes.

Torgerson evaluates the potential for an environmental discourse as an alternative to top-down bureaucratic rationalism. "Environmental management here involves 'discursive designs' such as participatory planning, regulatory negotiation, environmental mediation, and forms of public inquiry" (Torgerson 1990, 124). The difficulty lies in institutionalizing what is by its nature a transitory thing. While Torgerson thinks the possibilities of permanent patterns of administrative discourse are historically not good, I am a bit more optimistic. If environmentalism itself can be institutionalized through legislation and procedure, certainly discourse can be as well. The biggest procedural obstacles involve breaking elite domination of policy formation and focusing the discussion on addressing relevant problems (Torgerson 1990).

Kai Lee (1989) describes how environmental discourse and its numerous sustainability problems can be harnessed successfully in the Columbia River Basin study. Discourse is institutionalized in such a way as to involve four states, Washington, Oregon, Idaho, and Montana, their fish and game institutions, and regional Indian tribes, among others. Public input is a major component of this process. The Northwest Power Planning Council was established as the vehicle for policy debates by the Northwest Power Act of 1980. The problematic for discussion involved balancing the need for electricity, largely provided for by a series of hydroelectric dams, with the need to maintain a sustainable population of migratory salmon and steelhead trout. While the power companies suffered some additional costs through lost revenue, this discursive arrangement has facilitated solutions that are acceptable to the concerned parties (Lee 1989). Of course, this is heavily dependent on the availability of scarce water resources.

However, if discourse can be institutionalized to mediate interstate conflict, surely it can be applied to interlocal contamination scenarios such as military base reuse. This should be taken with a caveat, however, because it is often the case that local politics are the most contentious of all. Once discourse is agreed upon and established as an acceptable political process, the discussion can be focused on defining problems and assigning the appropriate solutions. By necessity, a democratic process requires decentralization not unlike that called for by Rivlin (1971). To be authentic, a variety of possible alternatives must be deliberated without prejudice. While governmental solutions are often appropriate, they are not necessarily the only ones possible. In cases where state and local institutional capacities are inadequate to mediate problems, private contractors can be employed. But if states and localities are better able to enforce environmental regulations than they were three decades ago, perhaps they can take the lead—along with certain environmentally friendly private interests—in the actual policy implementation, and leave the federal government to provide funding and guidance.

The big question is, can such arrangements work? And if they do, how would we know? In other words, how do we evaluate both the effectiveness of discourse and policy outcomes? If something goes wrong, who is accountable? These questions are not easy to answer without some systematic organizing scheme. There is a danger that evaluation methods can capture the discourse. Every effort should be made to ensure that this does not happen. However, if analytical assessments are agreed upon as part of the discourse as a way to measure progress on the policy agenda, then they are part and parcel of the discussion. Hence, methods do not capture the discourse, but are instead utilized to advance its cause.

The next chapter deals with how to evaluate the effectiveness of discourse in the implementation of economic development policy. A causal chain is illustrated between strategic planning, dealing with environmental externalities (both of which involve discourse), and development outcomes (job creation). The problem dealt with in chapter IV concerns the extent to which environmental damage blocks development.

Chapter IV

The Base Conversion Process and Developmental Variables

This chapter postulates the causal link between developmental discourse and policy outcomes, which are linked by the planning, funding, and implementation of complex base conversion plans. I discuss the three phases of base conversion and the practical steps that take place to reach the redevelopment goals of job creation and income replacement. Communities that develop planning organizations, and write grants, strategic plans, and environmental impact statements, within a short period of time may have an edge on those that do not, in terms of developmental success or failure. However, unless the intervening nature of environmental contamination and corresponding clean-up solutions are addressed, redevelopment goals will be much more difficult to attain. The data analysis attempts to get at these ideas.

The Pentagon mandates that all concerned local interests must be included in the base conversion process. To receive federal property, and planning and administrative grants, local governments must set up administrative structures that resemble what Charles Fox and Hugh Miller (1995) characterize as nascent forms of discourse. Many base conversions have resulted in success where these patterns of administration were properly developed. A discussion of how these structures are formed and function is necessary to more clearly understand how local governments plan, organize, and implement complex base redevelopment strategies, despite the confounding nature of pollution left behind by the military.

The conversion of military installations to private or mixed public use is one example of how brownfields redevelopment can be addressed and overcome. Types of pollution and strategies for remediation are evaluated with quantitative data in this chapter. Without addressing the military's leftover pollution, redevelopment is stopped dead in its tracks, for both practical and liability reasons. Other factors such as economic incentives are discussed for purposes of comparison. But the evidence suggests that environmental factors are just as important as the use of incentives, planning, and grant writing, for creating jobs and generating new income. This substantiates the notion that the local quality of life matters for development.

Opportunities for Redevelopment

Many communities view base closure with a sense of desperate fear for their economic well-being, due to the loss of jobs and federal income. However, base conversions often present substantial economic development opportunities that may eventually create more jobs and generate more income and tax dollars than were previously available through military spending, predictable though it may have been at one time. According to the Pentagon's Office of Economic Adjustment (OEA), "171,177 new jobs have more than replaced the loss of 87,557 DOD civilian jobs" (OEA 1993, 4) at ninety-seven former bases ordered closed between 1961 and 1993. This includes Vietnam and post-Vietnam era downsizing, as well as the first two post-Cold War Base Realignment and Closure rounds in 1988 and 1991. The OEA is an important source, because it is one of the principal grant-awarding agencies for military base conversion.

More than one-third of the bases closed prior to 1993 now are home to an educational facility of some sort, with four-year colleges, vocational/technical, and community colleges, being the most popular (OEA 1993). A recent example is the new California State University-Monterrey Bay, located on the former Fort Ord, which is located about ninety miles south of San Francisco (Associated Press 1995). Higher education is an important facet of economic development, not only because of short-term job creation and the influx of tax dollars, but also because of the long-term effects of having a skilled workforce that can command high-paying, value-added jobs. University-business partnerships are a good way to support technical innovation through research and development grants. Most redeveloped bases are home to some sort of industrial or office park, and almost half of the bases closed since 1961 are home to either a municipal or general aviation airport (OEA 1993). These types of infrastructure development are essential for the future of local economies as the relationships between cities become more competitive (see Peterson 1981).

Structural Factors

The transition from military missions to high-dollar private or mixed public-private developments is not an easy one, but it is possible provided certain procedures are carried out. The Pentagon requires that local civilians administer base conversions (OEA 1993, 4). Though the federal government and many states will provide redevelopment funds, the organization and implementation must be carried out at the local level. The federal requirement that reuse efforts be inclusive, as well as locally centered, contributes to the discursive tendencies of base conversion politics. The inclusion requirement is necessary to foster community consensus, to receive Pentagon recognition of the reuse effort, and to be eligible for Economic Development Agency (EDA) funding, as well as the transfer of former base property (Mayer 1992; *Federal Register* 1994, 740).

There are several reasons for local-level implementation. The Pentagon simply cannot afford the time and resources necessary for mediating interests, organizational development, and the implementation of reuse plans in all the communities affected by base closure. Also, as discussed in the previous chapters, federal control is often not practical. Many conversion delays are the result of foot-dragging or political haggling among federal agencies or specific branches of the military service. One of the greatest barriers to effective local implementation, according to results of the base conversion survey, was the failure of the federal government to get out of the way.

Local governments can be more efficient due to the decentralized nature of economic development, and because of their proximity to the situation. A key assumption of bottom-up implementation theory is that the closer one is to a problem, the more likely one will be to accurately assess it, provided there is enough street-level autonomy to do so (Lipski 1976; Hjern 1983; Elmore 1985). The old cliché, "If you want it done right, do it yourself," may be an apropos description of local base conversion success stories.

Base Retention and Conversion

Organization of the conversion effort usually takes the form of a successive evolution of three functional forms, which parallel the redevelopment process. These include a retention task force, a planning and reuse committee, and the Local Reuse Authority (LRA). The different circumstances that provoke the formation of these organizations will be examined here in greater detail.

Base conversion efforts actually begin to evolve before the base is anywhere near being closed. When the military begins its process of recommending bases for closure, local governments mobilize to prevent this from happening. The usual pattern is to create a retention task force, which is charged with lobbying the military, the BRAC Commission, and political leaders. These task forces engage in what is commonly known as a "dual track" strategy, which heavily emphasizes the retention effort while at the same time constructing the initial conversion plans. These plans are kept quiet until the BRAC appeals process is over and the final "hit list" is submitted to the president.

If the base is saved from the BRAC chopping block, the task force disappears, possibly to return in a few years during the next round of closures. But if the base is slated for closure, the retention committee persists past the closure/lobbying process and becomes the basis for the conversion planning and reuse committee (CPS 1995), and ultimately the LRA. It is at this time that local governments must educate themselves about the planning and base reuse processes.

Gaining Expertise

Most communities hire a consultant to assist with the planning and legitimation processes. Generally speaking, there is little base conversion expertise among local governments.[1] A consultant can pinpoint barriers to successful redevelopment and deal with the alphabet soup of federal agencies, which provide development grants and transition assistance. Many consultants are also experts at organizational development, helping to create an inclusive working relationship among diverse and opposing local interests.

Consultants play a crucial role in negotiating with the military for surplus and personal property, interim leases, and caretaker agreements. They are also responsible for marketing base facilities and soliciting private businesses to relocate as new tenants, which is essential for economic development (Morgan and Meyer 1995). The reason why consultants are more desirable to handle these functions than local officials is because many of the things that they do must be taken care of within the first few months after the base has been slated for closure. By the time local officials have learned how to handle property negotiations with the military, write OEA and EDA grants, and set up a Local Reuse Authority, they may have fallen behind in the process, and find themselves ineligible for many federal and state benefits. Therefore, the consultants are necessary to assist local officials in climbing the readjustment learning curve in relatively short order.

Reuse Planning and Organizational Development

Upon realization of the closure decision, the retention task force will evolve into the planning and reuse committee, and may, or may not, include many of the same individuals involved with closure prevention. As suggested by its name, this is where, and when, strategic planning occurs. To receive federal seed money for planning, the group must be representative of all factions of the community affected by the closure. This may include institutions, such as municipal, county, and sometimes state governments, school or special districts, and universities, or groups, such as regional government coalitions, special interest groups, chambers of commerce, racial minorities, homeless advocates, and industrial developers. It is important that planning groups be inclusive, because they are subsequently divided into subcommittees, and because they are the direct precursors to reuse authorities (*Federal Register* 1994).

Local Reuse Authorities are the officially recognized redevelopment entities that are sanctioned by the Department of Defense. The DOD sends administrative personnel out to the community to evaluate the inclusiveness of the planning and reuse group before it can be sanctioned as an LRA. This must be done to be eligible for OEA and EDA grant money or other benefits, such as land use or property acquisition. The Pentagon must approve the implementation plans prior to developmental grant funding. These plans must include a strategic plan for (at least) the first year of operation and job descriptions of the staff to be hired (*Federal Register* 1994; Morgan and Meyer 1995). In other words, local developmental discourse is evaluated and legitimized by the federal government.

Any local group or institution that feels it is being deliberately excluded from the planning committee can appeal to the DOD prior to the recognition of the LRA. However, to have an effect on the process, they must legitimately claim to represent an important sector of the community, such as a racial or ethnic group, the business community, or another governmental jurisdiction. If differences remain among the concerned parties, they must typically be resolved at the local level before the Pentagon will sanction the committee and LRA (*Federal Register* 1994). Though at first glance this may not seem to be terribly important, it can represent an impassable barrier to reuse planning and discursive policy implementation.

The Pentagon usually looks to the governmental jurisdiction where the former base property is located to head the planning committee and LRA. This function usually is granted to a city or county (Mayer 1992). The lead agency is generally the one charged with writing the redevelopment grants. Methods of local planning committee and LRA formation vary depending upon the types and numbers of actors involved. However, whether they are created through election or by appointment, most have the authority of local districts, which means that their existence is implicitly sanctioned by the states.

Direct legitimation is not required by states, although the smaller the state, the more likely that it will be directly affected, and hence involved in the conversion effort (see also Glassberg 1995). For example, the state of Louisiana was involved in the establishment of the England Authority in Alexandria, granting it the power to issue local bonds, the eligibility to receive state matching funds, and helping in the federal grant writing process (Morgan and Meyer 1995). Larger states, such as California, Florida, and Texas, were less involved in the redevelopment process, preferring instead to encourage local efforts. As they began to suffer the cumulative effects of statewide closures in successive rounds, larger states took a greater interest in assisting localities with base reuse. However, there were few substantive manifestations related to this concern. California discussed state income tax breaks for base reuse in that state, while Texas created a statewide consortium of public officials to assist communities, but neither state allocated many financial resources to these efforts. Nor did either state attempt to directly involve itself in conversion efforts the way smaller or medium-sized states such as Louisiana, Massachusetts, and New Hampshire did.

The size of planning committees and LRAs varies depending upon circumstances and location. Two good Texas examples include Carswell Air Force Base in Fort Worth and Chase Field Naval Air Station in Beeville.

The mayor's task force in Fort Worth originally included about 80 people on planning committees and subcommittees. This was an interlocal policy group, which included three cities and Tarrant County. They were responsible for the creation of the strategic plan and LRA bylaws. This group was eventually scaled down to a nine-member LRA. The local governments appointed the officials on the Authority, which included four from the county, three from Fort Worth, and one each from two other small cities in the immediate area. Each of these local entities had veto power over the LRA's activities (NAID 1994).

There were ten distinct political and business entities in the Beeville area, which were represented on its planning committee (the BBCRC), including the city, Bee and Goliad counties, four school districts, the water district, a government-appointed private industry council, and the chamber of commerce. There were twenty-three seats on the committee, which were allocated according to the size of the entity involved. The LRA represents the same institutions, with the exception of Goliad County, which was replaced by the Hispanic Chamber of Commerce, but was scaled down to fifteen members by 1995. Most LRAs do not have such a large membership—usually seven to nine individuals—since they are substantially scaled down versions of the planning committees. Because of the large size of the group, none of the members had veto power in the Beeville LRA bylaws (NAID 1994).

Once a planning committee is established and recognized by the Pentagon, an executive director and staff must be hired. These people are directly in charge of implementing the strategic plan and carrying out the day-to-day operations of the reuse effort. They also represent another good source of expertise, which can complement that of the reuse consultants. Ideally, some of the same people involved in the planning and legitimation processes, such as city and county staffers or the consultant, will be retained for the sake of continuity. Often the executive director is a former military officer who has served as a Base Transition Coordinator (BTC). A BTC is the military's community liaison during the closure process.[2]

It is essential at this time to appoint an environmental coordinator to oversee the degradation issues associated with the base and the redevelopment efforts. This is a permanent position because environmental remediation is a permanent concern of any reuse program, and is perhaps the most significant barrier to a successful conversion. It is also at this phase that the inclusion of private interests is crucial, since they will be the ones who take most of the risks associated with generating investment and creating jobs. Because base reuse is job-driven, implementation efforts must be focused on this goal. To this end, the business community cannot be excluded (Morgan and Meyer 1995). But by the same token, no one will do business on a contaminated site, so environmental issues must be dealt with first or the planning effort will stall.

Data and Method

Between 1988 and 1995, 115 major and minor military installations were recommended for closure or realignment by the independent BRAC commissions. There were a total of four closure rounds, spanning the administrations of three presidents, Reagan, Bush the elder, and Clinton. Now that several years have lapsed since the final round, we should be able to evaluate the effects of base closure and realignment on local communities. It appears that in many instances, base conversion has been a boon to local economies. This study attempts to assess why this happens, or conversely, why it does not in some cases.

Of the 115 BRAC cases, we received 44 surveys (roughly 38 percent) that we were able to use in this analysis. The executive directors or their representatives from each community affected by BRAC decisions completed the surveys. Due to the small size of the sample, many of the findings are descriptive as opposed to the usual level of methodological sophistication found in typical social science research. Nevertheless, these descriptions shed a great deal of light on the base conversion process. Even practitioners who are familiar with their own military transition might find some of the aggregate results interesting.

The sample contains cases from each BRAC round (1988, 1991, 1993, and 1995) and from all four branches of the armed forces. Since the Air Force took the hardest hit from the military downsizing process, it represents the largest group among the services in the sample. The sample also includes cases from all regions of the United States, from both rural and urban areas, from major and minor installations, and from large and small states. While California suffered the brunt of BRAC action (twenty-two closures or realignments), it is somewhat underrepresented in the sample. There is also considerable variation from those cases that report substantial success in the creation of new jobs and income and those that do not. As far as I can tell, this sample is very representative of the total population of base transitions, and hence the findings should be generalizable to this same population.

The survey attempts to identify specific actions in the base conversion process and tie them with tangible results. It also attempts to identify confounding factors in the process, such as contamination issues. Base conversion is a very difficult and time-consuming activity for most communities. There are many steps at which mistakes can be made, but a few errors or delays do not necessarily doom a community to high unemployment and lost income. Besides difficulties resulting from local infighting, state and federal agencies also cause developmental headaches for base transition efforts.

Many of the variables were measured in monthly increments, and hence are interval-level data. This was done for two reasons. First, it allows a benchmarking comparison of bases closed at different times by measuring the length of time between the closure decision and when a certain action is taken, or task is completed. Second, it extrapolates which steps in the process are likely to cause the most delays or setbacks. Other variables are descriptive (nominal-level data) and simply report what happened. Some of the dichotomous variables were recoded into indices to provide more variation. The variables that are likely to shed the greatest light on the nature of the base transition process, however, are those that measure grant money, jobs created, and income generated as a result of conversion efforts. The findings are presented in the form of tables for purposes of illustration. It is necessary to evaluate the descriptive data to find out what transpired before we can do a more sophisticated analysis regarding causality. The data were entered into SPSS 11.0 for Windows, which also helped create the tables.

Developmental Variables

The goal of most communities anticipating base conversion is economic development, usually in the form of skilled job creation, income generation, and subsequent multiplier effects. The means by which economic development is realized often include elements of strategic planning, grant writing, incentives, and the like, with combinations of different types of public and private reuse solutions as the intended outcomes. The process by which these solutions are realized is known as complex implementation. James Mercer, Susan Woolston, and William Donaldson (1981), Robert Stein (1990), and Mercer (1992) indicate that localities regularly engage in complex implementation to expand local infrastructure, economic productivity, and revenues. What is less clear, or often taken for granted, is the nature of the complex implementation process. This section addresses how to measure and evaluate the implementation of complex policy solutions at the local level.

Economic Development

The main goals of most economic development efforts are job creation and income generation, and military base conversions are no different. These are the dependent variables in this analysis. Economic development involves the effects of structural change on an economy, polity, or social system (Aharoni 1977; Frey 1983; Mercer 1992). Bruno Frey indicates that many societal and political factors are overlooked when it comes to analyzing economic activity. He argues that economic policies are made in a social framework, and that neither the economy nor the political system can be isolated from one another (1983). This is true because as political scientists, we generally accept David Easton's (1953) idea that politics is the allocation of scarce resources and values. Economics as a method of allocation is also important, because that is the way most goods and services are distributed in our society.[3] However, when resources are allocated authoritatively, the process is politicized. Addressing the complexities of base conversion is directly related to how structural changes are implemented within a social, historical, political, and economic framework.

This argument should not be confused with the corporate model advanced by Paul Peterson (1981). In fact, it is more in accord with Clarence Stone and Heywood Sanders's "regime paradigm" in which structural constraints are "mediated through the political arrangements that enable a prevailing coalition to govern a community" (1987, 16). It is impossible for cities or communities to operate only as economic actors, at least if they claim to represent the public interest. "[A] community cannot know in concrete terms what the public interest is, independent of political activity" (Stone and Sanders 1987, 15).

The sociology of a particular community will play an important role in determining the nature of its politics. Stein argues that in cities characterized by substantial socioeconomic cleavages and heterogeneity, politics is often adverse to the public interest. "Greater socioeconomic diversity provides greater opportunities for competing political parties to exploit the city's diversity by pandering to the demands of special interests" (Stein 1990, 27). Conversely, in smaller, more homogeneous communities, the reverse is likely to be true. This work makes the same assumptions as Clarence Stone, Heywood Sanders, Robert Stein, and others, that cities actively engage in economic development. The socioeconomic context, as well as the political context, will play a role in how local governments engage in the economic development activities of job creation and income generation, and determine who benefits from it.

Economic development is not the same as economic growth, although the terms are sometimes used interchangeably. Economic growth can be both a means to development and a measurable outcome. Growth is usually explained in terms of per capita earnings, or some other measure that denotes changes in income (Aharoni 1977). Economic development, however, is a somewhat more subjective concept. This is because it is based in part on the notion of progress. One society's notion of progress is often different from another's, resulting in differences of opinion with regard to the consequences, and externalities, associated with economic growth.

Current notions of progress in the United States seem to be preoccupied with organizational downsizing, the decentralization of authority, and privatization (see Rivlin 1971; Knott and Miller 1987; Osborne and Gaebler 1992). Base closings and conversions represent one small facet of this overall trend. The need for greater frugality, efficiency, and responsiveness seem to be the driving forces behind this tendency (Osborne and Gaebler 1992). However, the immediate effects of these practices may be more akin to economic recession than economic growth. Downsizing may result in greater levels of unemployment, or a temporary reduction in service performance, as reorganization takes place. This is usually the case when a military installation is closed (Dardia et al. 1996).

Without adequate preparation for the externalities of Pentagon downsizing, communities will experience higher rates of unemployment and negative economic growth. This is especially true for cases with intractable pollution problems. In rare instances, it may mean the death of an entire town. Such consequences can be averted or neutralized with appropriate economic development initiatives. In other words, developmental goals and subsequent growth can be achieved through the implementation of a strategic plan, which addresses the unique characteristics of a community and its military installation(s). This is not an unreasonable demand to place on local governments, because to a great extent, cities and counties are largely engaged in activities that are designed to enhance local economic performance. According to Stein (1990), cities actively pursue developmental policies that generate economic resources that can be used to benefit the community's economic welfare. In other words, they foster job creation.

The factors that determine whether a local government will succeed at redevelopment efforts are highly contextual. This is likely to be true for base conversions, not only for the previously elaborated reasons that effect localities, but also because of the differing levels of mission-related specialization, and existing infrastructure, that characterizes military bases. Despite such differences, developmental goals are generally similar across cases. Economic development is the main concept, which in this study is operationalized by the dependent variables of job creation, income generation, and the multiplier effects resulting from those activities. Of these, job creation is most likely to demonstrate the greatest effect, because there is greater lag time for the impact of income generation and multiplier effects.

Independent Variables

There are a number of independent and intervening variables that affect conversion outcomes. The problem involves sorting out which are the most important in terms of causality and which are important as intervening, or confounding, variables. I hypothesize that the key causal variables in the base conversion process are strategic planning and grant awards (in dollars), since those are the best indicators of discursive progress, as judged by the federal government. Other independent variables include the appointment of administrative positions, early completion of the environmental impact statement (EIS), and the use of economic incentives to attract new development. Key intervening variables involve the length of the military withdrawal (drawdown), the different types of environmental contamination, and the different types of clean-up strategies used to alleviate these problems, which will be discussed in greater detail below. The main hypothesized causal relationship looks something like the following:

Strategic Planning	→	Contamination	→	Job Creation
Grant Dollars		Restoration		
		Military Drawdown		

Strategic Planning

The process through which many communities make the transformation from dependence on military spending to economic development and diversification incorporates the basic elements of strategic planning. Whether a community achieves a successful base conversion may depend in part on the nature of its strategic plan. One of the objectives of this book is to ascertain how strategic plans are constructed at the local level. A key assumption is that strategic planning is an essential element of the complex implementation process and, as such, must be utilized by local governments attempting to undergo a base conversion. Perhaps more important theoretically, it is a product or outcome of administrative discourse, which is more difficult to empiricize.

According to Bonita Melcher and Harold Kerzner, strategic planning is "formulating and implementing decisions about an organization's future direction" (1988, 1). This preoccupation with the future is crucial for planners, as institutions tend to get caught up in the present-day realities of organizational and budgetary maintenance, using the past as a guide for operations. This kind of temporally based bureaucratic inertia makes it difficult for decision makers to adapt to an ever-changing environment. Melcher and Kerzner (1988) discuss the top-down nature of planning in the corporate sector, which gives us an insight as to its origins, but do not provide us with a model that is commensurate with addressing contemporary bottom-up public policy problems, such as base conversions.

Perhaps the foremost authority on public sector planning is James L. Mercer. He states that "[s]trategic planning is primarily a process rather than a set of specific products" (1992, 6). This process essentially involves the setting of goals, or at least a general direction in which one wishes to progress. Once particular goals have been identified, then a determination must be made as to how they can be attained.

Strategic planning originated within the armed services and became popular with large corporations in the private sector during the 1950s. It was essentially a top-down method of internal implementation designed to reduce risk, train managers, and develop consensus (at least among top-level administrators). "It begins with establishing aims and purposes, followed by formulation of ways and means to achieve those purposes, and provides direction for implementation of operational or tactical planning" (Mercer 1992, 19).

There are some important complexities in the implementation of strategic plans by local governments, which involve the local actors and institutional patterns of operation. These might include stronger grass-roots organizations, local elites who are opposed to certain types of development, different budgeting procedures, greater autonomy among local administrators, and innovative bottom-up management reforms. Strategic planning by local governments necessarily involves a more bottom-up approach than one would find at other levels of government due to the necessity of including multiple and diverse actors in the process.

Strategic planning is not a form of budgeting, although budgets are essential to any implementation endeavor. Nor is it particularly long-term in its focus. However, Mercer indicates that both budgeting and long-range goals can—and should be—aspects of a strategic plan. "An effective strategic planning process should be an important link between comprehensive planning and the annual operating and capital budgets" (Mercer 1992, 7).

Public-sector strategic plans are often short- to mid-term in their orientation. But military base conversions generally involve long-term commitments and have long-term implications for local governments, especially when one takes into account the recruitment of private enterprises, and other government agencies, for extended economic commitments in many transitions. These differences are reconciled, however, when one considers the reiterative process of strategic planning. Though original plans may be short-term in focus, they are often periodically updated to account for changes and externalities. This ensures continuity and the enhancement of efficiency and responsiveness. According to Mercer (1992), local government strategic planning must account for at least four elements. These are:

1. Evaluation of the total external and internal environment within which the community and the organization exist
2. Identification of major threats and opportunities facing the community
3. Assessment of the strengths and weaknesses of the organization and the community
4. Development of a concentrated strategy enabling the organization to effectively coordinate all community resources and operate within its environment (Mercer 1992, 6)

Corporate-sector strategic planning generally emphasizes only the internal and external factors. Clearly, Mercer's criteria are broader, more inclusive, and therefore more helpful to managers in the public sector. Another way to characterize such activities is through the acronym SWOT, which stands for strengths, weaknesses, opportunities and threats. This is a somewhat more simplistic view of Mercer's notion of assessing what should be done next, but is helpful for local discourse nonetheless. The SWOT method was used in the case of Lubbock, Texas, which oversaw the redevelopment of Reese Air Force Base (AFB). I was involved in this case as a participant-observer and can attest to the effectiveness of this process in maintaining a goal-oriented focus in the discussion. This case will be dealt with in greater detail in chapter V.

Can strategic plans fail? The answer is yes. They are not infallible. Just because managers agree on a strategic plan is itself no guarantee that they will achieve redevelopment success. Melcher and Kerzner (1988) indicate that there are several reasons why strategic plans often fail to attain developmental goals.

The first reason is that public officials neglect to reexamine implementation efforts. There must be continuous evaluation to fix mistakes and to eventually attain developmental goals. Second, managers can be guilty of extrapolation, or assuming that the future will be much like the past, when that may in fact not be the case. Third, planners can be blinded by initial success and grow complacent or overly optimistic. Finally, they can be guilty of encouraging overresponsiveness to current trends, resulting in overextension, and spreading organizational resources too thin (Melcher and Kerzner 1988, 13-14).

Grant Writing

For the most part, grant writing for base conversion is not a difficult proposition, provided that the proper procedures are carried out. Most cities have experienced grant writers on staff, so this task usually falls to whichever city (or sometimes county) is the lead agency in the conversion effort. In the event there is not sufficient expertise among local officials with the federal grant-writing process, base conversion consultants, such as NAID, can provide this service. Though most grant dollars come from the Pentagon's Office of Economic Adjustment and the federal Economic Development Agency, these are not the only venues that support and legitimize local base conversion efforts. For instance, private foundations also provide grants for defense readjustment (Morgan and Meyer 1995).

However, federal agencies remain the largest and most reliable source of seed money for base transition. According to the National League of Cities, community development is the best indication of federal financing from OEA and EDA (Mayer 1992). Other federal and state agencies provide money for transition endeavors as well. Base conversions that utilize former military airports as part or all of their transition plans are eligible for substantial aid from the Federal Aviation Administration (FAA). Those that suffer great numbers of civilian employee layoffs are often eligible for job retraining aid from the Department of Labor (GAO 1997).

Difficulties with transition grants can occur when there is no community consensus on strategic conversion plans. The federal government encourages localities to work out their differences amicably, but this does not always happen. In such cases, the military may assume control and simply sell off the former base property for whatever it can get for it, leaving the estranged parties in the lurch. But this option is a last resort and is not practiced except in the most extreme of circumstances.

Intervening Variables

Base conversion is a highly complex endeavor, in which any number of problems can manifest themselves depending on the nature of the particular case in question. Just to keep it interesting, there are any number of potential roadblocks that can derail or delay base redevelopment. These take the form of intervening variables in this analysis, including intergovernmental disputes, environmental degradation, and the length of military downsizing.

Intergovernmental Problems

The federal government causes most transition delays through unwieldy rules, regulations, and environmental degradation, but states can also get in the way of local conversion efforts. At times, local governments and local institutions squabble among themselves, which may even lead to litigation. Often these fights are simply disputes regarding the best potential use of the base property, but sometimes the motivation is simply greed, or an effort to expand bureaucratic turf relative to other cities or institutions. Federal roadblocks to development, such as lengthy military drawdowns, and excessive rules and regulations are all measured by the survey, as is state interference. Interlocal, interinstitutional, and litigation delays are also measured.

Environmental Degradation

An unanticipated consequence for communities located near military bases concerns the contamination caused by hazardous wastes, fuel, oil, ammunition, and the like. Much of the pollution was done at a time when people were largely unaware of the dangerous effects of the chemical contaminants to the ecosystem and local population. Some such as John Hird (1994) argue that the federal government is the largest polluter in the country, and that it actually makes it worse by delaying clean up through wasteful programs such as Superfund. And today, despite the greater knowledge of the harmful effects of certain chemicals, the military may continue to pollute and deny accountability (Jacob 1992).

To redevelop former military bases, the facilities must be cleaned up. The military is responsible for seeing that its contamination is neutralized and that the property is fit for industrial and (sometimes) residential use. The EPA has the power to approve reuse plans and environmental impact statements at former installations. But for the most part, the Pentagon and the states provide the greatest amount of aid for communities having to deal with military-related pollution.

The biggest questions concern which types of pollution are likely to cause the most transition delays. It is difficult to get at just how damaging the contamination is across cases in a study of this type. Therefore, it must be inferred based on the nature of the pollution and concomitant solutions involved. The case studies provided in chapter V shed more light on the severity of these problems. Nevertheless, the quantitative survey data can still help to substantiate the hypothesized causal order of the variables.

Length of Military Drawdown (downsizing)

Timelines are a way in which the progress on planning, organization, and implementation of base conversions can be compared and evaluated. The military utilizes timelines as a way to schedule drawdowns (equipment and weapons withdrawal and personnel reassignment) at bases that have been ordered closed or realigned. Drawdowns last from the time of the closure decision to the actual closure of the base facilities. A community's conversion timeline may last from the closure decision to the end of the implementation of strategic reuse plans. After this time, which varies depending on the community and the scope and duration of the plans, new plans and timelines can be formulated, or the previous ones can simply be updated.

A base that is destined to have its mission permanently closed is known as a "clean kill" in the closure/conversion practitioners' vernacular. These clean kills are usually completed within two to four years of the closure decision. Chase Field Naval Air Station (NAS) and Reese AFB in Beeville and Lubbock, Texas, respectively, England AFB in Alexandria, Louisiana, and George AFB in Victorville, California, are all examples of clean kills. Missions that are redirected or realigned with another base often take longer due to organizational and logistical complications. Examples of this latter type include Kelly AFB in San Antonio, Texas, and El Toro Marine Corps Air Station (MCAS) in El Toro, California, and NAS Cecil Field in Jacksonville, Florida.

The distinction is important because those communities that do not experience a clean kill have the luxury of formulating longer timelines to plan for the conversion or realignment. However, it seems that this luxury is often lost in certain instances, as localities lose sight of the prime conversion goals of job creation and income replacement. A timeline is not a substitute for inclusion, consensus, or organizational development. How a community progresses on an actual timeline may be an indication of the degree of success at its efforts to engage in discourse, strategic planning, and organizational development. Presumably, success at these preliminary activities will translate into successful redevelopment outcomes.

Substantive discourse is apparent in cases that demonstrate timely creation of planning committees and subcommittees, strategic reuse plans, the hiring of consultants, and reuse staff. Grant applications and awards indicate further evidence of base conversion progress. In cases that fail to achieve redevelopment goals, the blame can most likely be placed on a lack of authentic intergovernmental discourse, as there is little organizational development activity and few tangible policy outcomes resulting from planning discussions. As we shall see in subsequent sections of the book, this is due not only to the absence of discourse, but also to the practice of exclusion and inter-local turf wars, which often manifest themselves in the form of litigation, which is an expensive and unnecessary use of taxpayer dollars.

Policy Processes and Outcomes

This section analyzes the hypothesized relationships described above with data from the base conversion survey. While it is difficult to get at many variables that attempt to measure discourse (see Hansen 2003), strategic planning and grant writing variables are only realized with federal legitimation of local discourse. In an effort to make the data accessible to a wider audience, complicated methodologies are only used when necessary to demonstrate causality between the variables.

Dual Track Retention and Planning

As stated previously, conventional wisdom has it that local governments publicly fight base closure to preserve the economic benefits and prestige associated with having an installation nearby, but that they privately plan for conversion in the event that the retention effort is unsuccessful. Perhaps it is the case that communities affected by closure decisions did not fight BRAC hard enough, or aimed their efforts at the wrong venues. Most local officials would probably disagree with that assertion however. It may be that considerable effort was placed on base retention, and that only when closure or realignment was imminent (following the announcement of BRAC decisions) did communities begin strategic planning and organizational development. The evidence suggests that despite protestations to the contrary, little more than a third of the communities that lost their bases truly pulled out all the stops in the retention effort.

Only 37 percent of the surveyed communities afflicted with base closure or realignment chose to organize rallies during BRAC visits or during the time when preliminary closure lists were being drawn up by the individual armed services (see table 4.1). Of those municipalities that did organize public demonstrations, only a small percentage of the population typically participated. The participation level of the subset that reported public demonstrations is not large enough to report. But public rallies or other "outside strategies" are not the only way in which a community can make its case for base retention.

Another 37 percent reported spending money on "inside strategies" such as lobbying the armed services, elected officials, and BRAC commissioners on behalf of their installations (see table 4.1). While about 60 percent of the surveyed cities did not spend anything on base retention, the remainder spent anywhere from a few thousand up to $500,000. The mean dollar amount spent on base retention among the cases surveyed was $67,611.14, which really is not very much.

Table 4.1 Local Base Retention Activities

	Frequency	Percentage
Public Support (rallies, demonstrations)	16	37
Spent Money on Base Retention	16	37
Hired a Lobbyist or Consultant	15	35
Conversion Plans Made Prior to Closure Decision	15	35
Inclusion of Local Business Community	28	67

N = 44 Note: Respondents may pick more than one category.

Thirty-five percent reported hiring a lobbyist or defense consultant to avert base closure (see Table 4.1). Presumably, that is where much of the retention money was spent. Interestingly, a preclosure step has been added to the BRAC process for 2005 that is designed to thwart the hiring of such consultants in order for local governments to avoid this expense. However, it does not appear that this was an unbearable cost for those communities affected by closure. Of course, I did not survey those communities that successfully averted base closure. Perhaps part of the reason why the cities in the base conversion survey suffered base closure is because they did not put their money where their mouths were when it came to base retention. I will deal with related issues below.

Thirty-five percent of the respondents indicated that they had made tentative conversion plans prior to the base closure decision. Whether they made reuse plans or not, 67 percent of the cases reported involving the business community in the retention effort (see table 4.1). Involvement of business interests, whether it is formal inclusion of the local chamber of commerce in a lobbying effort or more of an informal solicitation for base transition alternatives, indicates some level of preconversion planning and organizational development. The quantitative data seem to bear out the assertion that the dual-track retention and planning process is an authentic, community-oriented, contingency activity. What is less clear is if the retention efforts were merely for show to prepare local citizens for the task of base conversion. While several were probably serious efforts to preserve the status quo, most were probably half-hearted, as indicated by the evidence.

Representation on House or Senate Armed Services Committees

Naturally, a community would lean on its congressional representatives when faced with base closure, despite the independent nature of the BRAC process. But was this an effective strategy? Sixty-nine percent of the respondents indicated that their communities did not have political representation on the U.S House or Senate armed services committees, nor did any of their elected federal officials hold a leadership position in either of the two major political parties (see table 4.2). But 31 percent did have such representation—sometimes in both categories—though it was not enough to prevent closure or realignment (see table 4.3).

Table 4.2 Congressional Representation (U.S. House and Senate)

	Frequency	Percentage
Member on House or Senate Armed Services Committees	13	31
Member in Leadership position (of either political party)	13	31

N = 44

This suggests that the BRAC process was somewhat successful at removing congressional politics from base closings as intended. But without a comparison to those bases that remained open it is difficult to say for sure. Those communities that had such representation in Congress surely must have had their high hopes for base retention dashed by BRAC.

There is anecdotal evidence to suggest that bases in the districts of Representatives Charlie Bennett (D-FL), Sonny Montgomery (D-MS), and Jamie Whitten (D-MS) were saved due to the positions of power these men occupied in the House, at least in the early rounds. This would be an indication of the second face of power (Bachrach and Baratz 1962). But there were others, such as Strom Thurmond (R-SC) and John Warner (R-VA) in the Senate, who were equally powerful but whose presence did not affect BRAC decision making. Perhaps their status as members of the minority party in Congress explains this adverse effect.

Table 4.3 indicates that almost 17 percent of the cases surveyed had clout with both the relevant defense committees and party leaderships in Congress. However, they still lost their bases. Looking at those cases that were represented by one or the other, there did not seem to be a substantial difference between whether they were represented on the committees or by the party leadership; neither one helped retain their bases.

Not surprisingly, a greater number—roughly 55 percent—were not well connected to the leadership or the relevant committees in either house of Congress. The moderate, but significant, Cramer's V measure of association does indicate that at least statistically there is something to the second face of power argument among the cases surveyed. So, while Congress was somewhat neutralized by the BRAC process, there were still some measurable effects of its involvement.

Table 4.3 Cross-tabulation of Community Congressional Representation

		Member in Leadership (of either party)		
		Yes	No	Total
Member on House or Senate Armed Services Committee	Yes	7 16.7%	6 14.3%	13 31%
	No	6 14.3%	23 54.8%	29 69%
	Total	13 31%	29 69%	42 100%

Cramer's V = .332 (Significant at the .05 probability level.)
N = 44

Transition Planning

Once the initial shock of the closure decision has sunk into the community's consciousness, the preliminary strategic plans are dusted off and organizational development begins. Transition planning can be measured across cases by timelines. Timelines are utilized by the military to coordinate its drawdown and closure processes with the community's strategic planning process. The nature of the mission at a base can dictate both its drawdown and the nature of the strategic planning process. Many closed bases were home to only a single mission, which made conversion a relatively straightforward task in many cases. However, many other cases have been characterized by rapid strategic planning and slow military drawdowns, which negatively affects localities. This can happen for a variety of reasons unrelated to the original military mission at the base.

As indicated in table 4.4, much of the early planning and organizational development tasks occur within a year for most communities. The median indicates that in most places this occurs relatively quickly. The mean exposes several outliers in the planning and organizational process. This is an indication of possible failure at community development, and interlocal or institutional infighting. Most strategic plans are completed within two years and approved by relevant federal agencies within one year of submission, for a total of three years.

While strategic planning is relatively evident in table 4.4 by completion and approval of transition plans and environmental impact statements, organization development is a little harder to come by. However, the rapid inclusion of the local business community, the hiring of relevant implementing personnel, such as the BTC, BEC, consultants, and the executive director, and the creation of the Local Reuse Authority all indicate that critical planning organizations have been formed.

Table 4.4 Base Conversion Planning (in monthly increments)

	Mean	Median	N
Begin Strategic Planning	4.76	2	41
Community Contacted OEA and EDA	3.83	3	41
Formal Inclusion of Local Business Community	4.5	2	41
Hiring of Base Transition Coordinator	9.2	3	31
Hiring of Base Environmental Coordinator	7.1	3	31
Hiring of Land Use Consultants	13.4	9	36
Hiring of Executive Director	13.9	4.5	40
Completed Strategic Plans	24	22.5	38
Completed Property Screening	12	6	38
Completed Homeless Screening	16	12	36
Completed Environmental Impact Statement	16.6	7	40
Completed Interim Leases	29	18	35
Received HUD Approval of Strategic Plan	32.6	29	28
Received OEA Approval of Strategic Plan	31.3	28	35
Received EDA Approval of Strategic Plan	25.7	24	12
Creation of Local Reuse Authority	21.3	10	40

The negotiation of interim leases indicates that communities are taking over routine maintenance functions at the base properties, such as security, grounds keeping, and the like, as the military pulls out. Interim leases are also negotiated with new tenants, who can sometimes move onto the property as the drawdown continues. When this occurs, the redevelopment effort acts as agents on behalf of the military, since the property is not officially transferred to the LRA until the drawdown is complete.

Base Conversion Solutions

Those who are familiar with the base conversion or realignment processes know they are very time consuming and often difficult. For those cases that were part of the early BRAC rounds (1988 and 1991), the military drawdown process was long and cumbersome. On the other hand, by being first, many of these cases benefited from the availability of greater amounts of grant money and recruitment of prospective tenants to occupy former military facilities. Cases from the 1993 and 1995 rounds typically had faster drawdowns, but had to compete harder for grant money and market share for tenants.

There are several categories of military base conversion, as illustrated by table 4.5. A single private solution refers either to privatization-in-place or to the relocation of a single private contractor to utilize the property. Privatization-in-place is an attempt to maintain preclosure employment levels and a similar mission, but with private instead of military contractors. This option has been taken away for the 2005 BRAC round because it is viewed as an attempt to "un-BRAC" the independent commission's closure decisions (DOD 2001).

Table 4.5 Type of Base Transition Plan

	Frequency	Percentage
Federal Transition	5	12
Single Public Solution	7	17
Single Private Solution	7	17
Multiple Use Solution	22	54
Totals	41	100

A single public solution is when one governmental entity takes over administration of the former base. These solutions typically involve the creation of an airport or university, although these may also be found as elements of multiple solution cases. Airports are usually funded with multimillion dollar FAA grants, while public universities are largely funded and administered by the states. A federal transition is one where another federal agency steps in and assumes control over the former base property. Base realignments are included in this category, as well as those cases which appear to be examples where the military has attempted to reverse the decisions of the independent commission by placing new missions at bases it would like to see remain open. Some of these will be discussed in the chapter V case studies.

The most popular option chosen by communities to resolve their closure dilemmas is multiple use, not privatization, as public choice theorists might expect. Typically, multiple use solutions involve having multiple tenants occupy the former base property. This can include both public agencies and private corporations. Such diversity often aids communities in the event that a disaster befalls one sector of the economy. For many communities, industrial parks of this sort are a favorite solution. As indicated in table 4.5, 54 percent chose a multiple use solution, while 17 percent each chose a single public or single private solution. Roughly 12 percent were co-opted into a federal transition or were realigned.

Types of Reuse Authorities

As mentioned above, most base conversions, including some federal transitions and realignments, are administered through Local Reuse Authorities. There is not one best way to do this, but due to the nature of local control regarding transitions, the most popular form of LRA is a city, county, or interlocal commission. Typically, a city or county will play the role of lead agency and have a plurality of seats on the board, and perhaps even veto power over group decisions. They may also be responsible for grant writing and hiring staff for the reuse effort once the initial plans have been approved. This represents a certain amount of risk on their part regarding the outcomes.

Table 4.6 Type of Local Reuse Authority

	Frequency	Percentage
City, County, or Interlocal Agency (including Enterprise Funds)	26	62
Special District/State Agency	12	29
Nonprofit Organization	4	9
Totals	42	100

According to table 4.6, 62 percent—almost two-thirds—chose the interlocal arrangement. This is not surprising, since it is the structure recommended by the Pentagon during the strategic planning process. Twenty-nine percent of LRAs are special districts or state agencies, which bestows an air of sovereignty that may not be present at the local level in all states. Ten percent are nonprofit organizations of some kind, such as local foundations, or trusts, that are set up to administer the facility in the public interest.

Many of the interlocal type of LRAs turn the administration of the facility over to a port authority, which is essentially a public enterprise fund. Many of the special districts are empowered to issue bonds or have other means of financing, which are similar to municipal enterprises, so the dynamic of both of these types of LRAs is often similar to that of government-owned corporations. This is what was done with part of the former NAS Cecil Field in Jacksonville, Florida, which is largely overseen by Jaxport, a local public enterprise that runs the local air and seaports (see chapter V for a more detailed discussion). Such operations typically have more experience with economic incentives used to attract prospective tenants or business partners, since they may already oversee enterprise zones or free ports.

Local reuse authorities are usually the ones responsible for hiring the executive director and staff, as well as land use consultants and planners. They are charged with formulating the official long-term strategic plans, environmental impact statements, and marketing of the base facilities.[4] Pursuant to this they decide which economic development incentives are appropriate for attracting the best tenants to the base in accordance with the strategic plan. LRAs may also negotiate interim leases with tenants and caretaker agreements with the military, and oversee the acquisition of the base when the military drawdown process is complete. They negotiate for real and personal property on the base, as well as the land and facilities themselves. Generally, they try to get the best deal possible for the installation, and attempt to recruit the tenants that will best add value to the property, so as to enhance the tax base. As Knott and Miller (1987) argue, the rules determine the structure, as well as who wins and loses, from different institutional arrangements. So, the structure of LRAs can potentially shape the nature of the conversion effort as much as a strategic plan. Ostensibly, community leaders have some notion of the most effective organizational structure needed to carry out their redevelopment plans. But like strategic plans, the organizational structure is subject to political negotiation, and is an outcome of local discourse.

Economic Incentives

Many base conversions provide a variety of economic development incentives to attract new businesses to their localities. These include tax increment financing (TIFs) and abatements, the establishment of enterprise or economic development zones, free ports, free or low rents, direct cash incentives, or services in-kind. These incentives are designed to reduce transaction costs, such as tax burdens or tariffs, which increase the costs of doing business. It is unclear from the literature whether these incentives make a great deal of difference, and almost 32 percent of the cases surveyed reported no use of economic incentives at all (see table 4.7). Presumably,

Table 4.7 Economic Incentives for Base Conversion

	Frequency	Percentage
Tax Increment Financing (TIFs) Or other Tax Abatements	14	34
Enterprise, Economic Development, Or Freeport Zones	18	41
Reduced or Free Rent	21	51
Direct Cash Incentives	3	7.3
Services in kind	13	31.7
No Incentives Provided	13	31.7

N = 41 Note: Respondents may pick more than one category.

these would mostly be the federal transitions and single public solution types of conversions.

Eighteen percent of the cases reported using only one kind of incentive to recruit new tenants. Twelve percent used two types of incentives, and roughly a third reported using a combination of three or more economic development incentives to drum up business. As seen in table 4.7, a third of the cases reported utilization of TIFs or related types of tax abatements. Forty-one percent created enterprise, economic development, or free port zones so as to avoid things such as inventory taxes or other transaction costs. Over half the cases provided space rent free or at reduced rates for their tenants. Less than 10 percent provided direct cash incentives or subsidies, but almost 32 percent provided some type of services-in-kind.

Previous studies on base conversion emphasize the difficulties of measuring the economic impact on local communities (Glassberg 1995; Dardia et al. 1996). This can be a problem because frequently military bases are isolated from local economies unless they have large numbers of civilian employees working at the facility (Sorenson 1998). However, it can be inferred that some base conversion efforts make a major attempt to integrate former military facilities into the communities at large, by placing them within new or existing economic development, free port, or enterprise zones. Whether successful or not, the attempt indicates a community-oriented focus with regard to base redevelopment.

Property Transfers

There are several ways in which LRAs can attain control over former military installations. They are Economic Development Conveyance (EDC),[5] Public Benefit Conveyance (PBC), property reversion, and purchase at below market value. Some communities never gain control of the former base property, however. This peace dividend was sacrificed so localities could survive the effects of base closure with a minimum of economic disruption. The costs of environmental cleanup also quickly minimized whatever financial benefits selling the properties might have generated.

Table 4.8 Method of Former Base Property Acquisition

	Frequency	Percentage
Economic Development Conveyance	28	68.3
Public Benefit Conveyance	7	17
Property Reversion	2	5
Purchased Below Market Value	1	2.4
Will Not Acquire Property	3	7.3
Totals	41	100

In a very small number of instances the federal government is able to get below market price for the property, but for the most part it gives the installations to the communities through several different transfer mechanisms. The predominant means for the intergovernmental transfer of military bases is what is known as an Economic Development Conveyance. As indicated in Table 4.8, more than two-thirds of the cases in this study received this type of property transfer. Another, similar method is called a Public Benefit Conveyance, which may or may not be done for the sake of economic development. Seventeen percent received property through this transfer method. About 7 percent of the surveyed communities will not acquire former base property. These are presumably the federal transitions and "un-BRACed" facilities that remain in the hands of the Pentagon or another federal agency.

A way to centralize the local conversion process is to have the property revert to the city in which the former base is located. For those communities that donated land to the armed services with reversionary clauses in their deeds during the 1940s, there is the advantage of having to deal with less bureaucracy, both at the local and federal levels. This is not to say that reversionary efforts are exempt from the inclusion/discourse requirement, only that the number of groups that claim to have a direct interest is decreased, and the city is exempt from certain federal requirements.

Property reversion is the optimal, if rare, type of transfer for communities. Reversion helps reduce federal delays because it frees localities from having to deal with the DOD's land screening process, homeless and American Indian claims, and Housing and Urban Development (HUD) approval of redevelopment plans. Some of these delays result from the federal government's effort to receive fair market value for its property in order to reduce its costs. Others are designed to assist disadvantaged groups, such as the provisions of the McKinney Act of 1993, which are aimed at addressing the homeless problem by providing low-cost housing and redressing American Indian tribal land grievances. The involvement of HUD is to ensure that these procedures are carried out in a fair and equitable fashion.

Reversion does not exempt localities from having redevelopment plans approved by EPA or DOD, nor does it remove the burden of responsibility for environmental cleanup from the federal government. However, it is not very common because many communities that transferred land to the federal government for the purpose of building military bases assumed the facilities would be permanent, and did not include reversionary clauses in the original deeds. As indicated in table 4.8, only 5 percent of the surveyed cases were able to use property reversion as a method of transfer. The only lesson that can be learned from this is that if cities or counties transfer land to the federal government for similar purposes in the future—say for homeland defense, drug enforcement policy, or other law enforcement—they need to make sure it can be legally transferred back.

Redevelopment Problems and Delays

As indicated in table 4.9, fully 77 percent of all base conversions sampled reported some kind of federal interference or delays during their transition. Twenty-five percent reported interference or delays from their state governments. About 45 percent reported that environmental contamination on the base significantly delayed their conversion efforts. A third indicated delays related to interlocal infighting. Another 36 percent reported institutional battles. Given the fact that most LRAs are city, county, or interlocal organizations, the latter two findings should not come as much of a surprise. Interlocal cooperation is going to experience growing pains. Such is the unintended consequence of democracy.

Almost 20 percent of the conversion efforts were involved in litigation with competing governments. This is disappointing because it is not a good use of taxpayer dollars. It is a major indication of a lack of local discourse and of hidden agendas. Certainly this is more wasteful for localities than the hiring of consultants in the retention effort, but the federal government has taken no steps to try to prevent litigation, other than to encourage local discourse. In rare instances, the military may divide the land and sell it to private interests for whatever it can get when localities cannot come to an agreement regarding reuse plans. The Navy recently threatened this potential outcome in the case of the El Toro MCAS in Orange

County, California. There were no cases where there was not at least one of these barriers to the implementation of local strategic plans. These findings indicate that intergovernmental conflict is, unfortunately, alive and well in our federal system.

Table 4.9 Sources of Base Conversion Delays

	Frequency	Percentage
Federal Interference or Delays	33	77
State Interference or Delays	11	25
Environmental Degradation	20	46
Interlocal Infighting	15	34
Institutional Competition or "Turf Wars"	16	36
Litigation among Interested Parties	8	18

N = 44 Note: Respondents may pick more than one category.

Pollution at Former Military Installations

The pertinent question for this section is to what extent pollution contributes to military base conversion delays. As mentioned above, almost half of the cases in this study attribute some delay to environmental problems. While it did not cause as many delays as federal interference, degradation has the greatest potential for long-term liability and negative effects on quality of life. It can also represent a big negative when it comes to marketing base facilities to prospective tenants.

Contamination at former military bases can take several forms. As demonstrated in table 4.10, 74 percent of the respondents indicated that their base had some form of chemical contamination, such as TCE, PCB, or other cleaning solvents. These are dangerous because they can pollute local groundwater supplies and affect wildlife, particularly birds that have the misfortune to come into contact with them. About 29 percent reported that they have live or spent ordnance of some sort on their bases, such as shells or bullets.

Seventy-four percent reported that there was oil or fuel stored or spilled somewhere on the base property. Typically this is located at former motor pools, but it can also infiltrate the groundwater due to runoff. Fifty-five percent of respondents reported that there were toxic or hazardous wastes on the base. This last problem is the stickiest, because it is the most expensive and potentially troublesome to clean up, and has the most potential to adversely affect conversion plans that involve attracting new tenants to former base property. Such cases can become Superfund sites.

Table 4.10 Types of Pollution at Former Military Bases

	Frequency	Percentage
Chemical Contaminants	31	74
Spent and Live Munitions	12	29
Oil and Fuel Contaminants	31	74
Toxic and/or Hazardous Wastes	23	55

Note: Respondents may pick more than one category.

Table 4.11 Cross-tabulation of Delays Caused by Chemical Contaminants

		Chemical Contamination		
		Yes	No	Total
Environmental Degradation Delays	Yes	18 43%	2 5%	20 48%
	No	13 21%	9 31%	22 52%
	Total	31 74%	11 26%	42 100%

Cramer's V = .351 (Significant at the .05 probability level.)

Several cross-tabulations of conversion delays and the different types of pollution were conducted to see which caused the most trouble. Some of the findings are significant, which would seem to indicate problem severity, but others are not. The question is, why?

Chemical contaminants (see table 4.11) seem to have caused the most significant delays, as indicated by the strong Cramer's V score of .351. This is especially important for former Air Force bases and naval air stations. Most of these chemical contaminants, especially TCE, were used to clean aircraft or sea craft. They are dangerous to water quality, and especially to groundwater reserves, though they are easily dealt with if treated with new and innovative technologies. This will be discussed in greater detail in the case studies presented in chapter V.

Spent or live ammunition could potentially delay conversion efforts, but generally does not. The findings of table 4.12, though interesting, are not statistically significant. This is probably because munitions pollution is not likely to be sited near human habitations or office space. As such, it can be parceled out and either cleaned up or just left alone.

Table 4.12 Cross-tabulation of Delays Caused by Munitions

		Spent and Live Munitions Pollution		
		Yes	No	Total
Environmental Degradation Delays	Yes	8 19%	12 28.6%	20 47.6%
	No	4 9.5%	18 42.9%	22 52.4%
	Total	12 71.4%	30 28.6%	42 100%

Cramer's V = .241

Table 4.13 Cross-tabulation of Delays Caused by Oil and Fuel Pollution

		Oil and Fuel Contamination		
		Yes	No	Total
Environmental Degradation Delays	Yes	7 40.5%	3 7.1%	20 47.6%
	No	14 33.3%	8 19%	22 52.4%
	Total	31 73.8%	11 26.2%	42 100%

Cramer's V = .243

Many bases with this type of pollution simply turn the contaminated areas, which are typically wooded or otherwise undeveloped, into wildlife preserves. If these areas are judged to be not terribly hazardous or unsafe for general use, they can be set aside for recreational purposes. Hiking or biking trails, wildlife educational-learning centers, and the like can serve as uses for some of these areas. Sometimes the federal or state governments will become tenants on the former base property and use such areas as training grounds for reserve units, or for the National Guard. Rifle or shooting ranges also can be developed here, depending on the nature of the overall strategic plan.

Oil and fuel pollution is cited as a noticeable cause of developmental delay by more than 40 percent of the respondents, as reported in table 4.13. But this finding is not statistically significant, despite the moderate Cramer's V score of .243. While oil and fuel can potentially affect wildlife and groundwater in ways similar to chemical contaminants, perhaps there was not as much of it as the chemical pollution. Or perhaps it was more easily cleaned up. If much of it was on the surface, as opposed to many chemical contaminants that have leached into the ground, this is probably the case. More likely, it was probably centralized in areas such as the motor pool, which are easily parceled away from the rest of the useable property. This is often not the case with other types of chemical contamination, which are dangerous because of unrestricted spread.

Table 4.14 Cross-tabulation of Delays Caused by Toxic and Hazardous Wastes

		Toxic and/or Hazardous Materials		
		Yes	No	Total
Environmental Degradation Delays	Yes	14 33.3%	6 14.3%	20 47.6%
	No	9 21.4%	13 31%	22 52.4%
	Total	23 45.2%	19 54.8%	42 100%

Cramer's V = .292 (Significant at the .05 probability level.)

Table 4.14 demonstrates one of the most damaging types of contamination present at military bases—toxic waste. Forty-five percent of the base conversion cases reported having this problem. This is a scary proposition for transition efforts because it is a potentially significant source of delay. Tenants do not want to move onto a facility contaminated with hazardous materials, both for safety and liability reasons. Citizens do not want to live near such facilities for health reasons. Without federal aid, communities that inherit property contaminated with hazardous wastes would without a doubt be dead in the water when it came to economic development. However, until the military has lived up to its statutory obligations to clean up its mess, the property in question cannot be transferred for civilian use.

Only a third of the total number of cases surveyed reported that hazardous materials caused them delays, which is reflected by the moderate level of the measure of association (Cramer's V = .292). This finding is statistically significant at the .05 probability level. An astute reader might ask why chemical contaminants were more troublesome for base conversions than toxic waste according to the measures of association. A guess is that it may have something to do with the severity and containment of the problems. Much like oil and fuel, toxic and hazardous materials are likely to be concentrated in certain areas of the base, while chemical contaminants can leach out of control.

Table 4.15 allows a comparison between the different types of pollution with regard to respondents' views of contamination-related developmental delays. As demonstrated in the previously discussed cross-tabulations, the Pearson's r correlations substantiate that chemical and hazardous material contamination are the most responsible for delays attributed to environmental degradation. The Pearson's r measures of association are roughly the same as the Cramer's V in the cross-tabs, and have the same level of significance.

Table 4.15 Correlations between Types of Pollution and Delays

	Chemicals	Munitions	Oil & Fuel	Haz-Mat
Environmental Degradation Delays	.351*	.241	.243	.292*

N = 42
* = Pearson's r correlation is significant at the .05 probability level (1 tailed test).

Table 4.16 Environmental Solutions at Former Military Bases

	Frequency	Percentage
Superfund Process (CERCLA and SARA)	15	36
Parcelization of Base Property	24	57
BRAC Clean-up Teams	36	88

N = 42 Note: Respondents may pick more than one category.

The data seem to substantiate the notion that, if ignored, environmental problems can significantly delay economic development. But when this happens, how do governments deal with the problems so that they can move on with their strategic plans? There are three main types of solutions employed—often in concert—to deal with military-related contamination. They are the Superfund, parceling the property into redevelopment and clean-up zones, known as parcelization, and the utilization of the military Fast-track or BRAC clean-up teams, which are an intergovernmental solution centered at the local level.

Well over a third of the respondents reported that they had Superfund sites on their bases, which is not surprising, given the number of federal facilities on the National Priority List (NPL). As a response to localized environmental degradation,

57 percent of the cases parceled their bases during the conversion process. Parcelization involves dividing the property into sections, which can then be zoned for different purposes. This practice was made possible by Clinton administration reforms to the base conversion process that were learned from mistakes made in previous rounds. The presence of NPL sites on a base can be neutralized, at least in part, by parcelization.

Table 4.17 Correlations between Environmental Problems and Solutions

	Parcelization	Superfund	BRAC Clean-up
Chemical Contaminants	.250*	.444**	.111
Munitions	.122	-.031	.057
Oil and Fuel Contaminants	-.078	.331*	.309*
Toxic and Hazardous Materials	.083	.378*	.271*

N = 42
** = Pearson's r correlation is significant at the .01 probability level (1-tailed test).
* = Pearson's r correlation is significant at the .05 probability level (1-tailed test).

Almost 90 percent of the cases utilized the BRAC clean-up plan, which is funded by the Pentagon and administered by the LRA, usually with help from the state environmental regulatory agencies or private contractors. The BRAC clean-up model most closely resembles the environmental discourse model discussed in chapter III, which utilizes elements of decentralization, privatization, and local control. BRAC clean-up teams are headed by the base conversion effort under the guidance of the Base Environmental Coordinator (BEC). Clean-up efforts are paid for by the Department of Defense, which is liable under several federal statutes for its pollution. However, local redevelopment efforts may not have the institutional capacity to engage in their own remediation, so they typically ask for help from the states and/or private contractors. This process will be discussed in substantially greater detail in the case studies presented in chapter V.

How do local base redevelopment efforts decide which solutions are best for their particular problems? The answer may have to do with severity. While no one wants a Superfund site on his or her property because of the associated bad publicity, localities may not have much choice under federal statute if they want to avoid liability. However, there are ways to ameliorate this problem. Table 4.17 correlates environmental problems at former military bases with concomitant solutions to see which match up together.

Superfund is the most common solution to deal with the severity of chemical contamination as indicated by the highly significant Pearson's r of .444**. But it also is apparent that the NPL sites are sometimes parceled out from the rest of the property as indicated by the statistically significant score of .250*. This can minimize the negative effects of having a Superfund site on the property. Fans of the Superfund/NPL process can take solace from the finding that it is the most statistically significant of the available solutions. Given the negative score of -.031, it is apparent that munitions pollution is not dealt with by the NPL system. Usually parcelization is sufficient to deal with this problem (r = .122), but it was not particularly strong, or statistically significant.

Both Superfund (r = .331*) and the BRAC clean-up teams (r = .309*) were used to deal with oil and fuel contamination. Again, the difference between when to use which solution probably depends on severity. For particularly intractable amounts of oil and fuel contamination, or when it is present with other types of chemical pollution, Superfund is probably the only option. However, when it is present in smaller amounts or is localized, the BRAC clean-up teams are probably sufficient to deal with the problem. Interestingly enough, parcelization did not seem to be an option with oil and fuel contamination. Perhaps it is the case that such sites do not need to be rezoned for restoration to take place.

When it comes to cleaning up toxic or hazardous materials, Superfund (r = .378*) and BRAC clean-up teams (r = .271*) are both utilized. Often when BRAC clean-up solutions are chosen, private contractors or state agencies are utilized to deal with the problem, with the federal government paying the bill. This is done in the hope that delays associated with being put on the NPL can be avoided. It is also potentially less expensive for the military. The idea for LRAs is to get the pollution problems cleaned up as quickly as possible, so that they can move on to the greater goals of economic development.

Policy Impact

While there is much anecdotal evidence from case studies to suggest that base conversion works to replace lost jobs and income, the results are still inconclusive. A 1998 GAO report argues that while base conversion is a necessary activity, the success rates of the different communities may be more of an artifact of the success of the national economy, rather than local development. If we measure base conversion success by jobs created, income generated, and the multiplier effect, compared to those rates at the bases before they were closed or realigned, we may be in for some sobering results. However, we must keep in mind that base conversion is a long-term endeavor, so short- or mid-term results are not the last word. The results after twenty years, instead of five or ten, might be more relevant.

In table 4.18, the column on the left represents the mean number of jobs lost (in thousands), while the column on the right represents the mean jobs created (in thousands) as a result of base conversion. The mean number of jobs lost at each base was 4,865, while the mean number created was only 896, as of January 2000. Keep in mind that many base conversions are still taking place or just recently completed, and not all the bases slated for closure had actually done so by January 2000. It is also important to remember the barriers to conversion that many LRAs continue to experience. Federal, state, and interlocal interference, environmental clean up, as well as competition for new tenants exacerbate this problem. I expect the number of jobs created to continue to rise in the near future as more of the bases actually finish closing, and as the new tenants continue to hire more employees.

The same goes for the payroll and multiplier effects of defense spending in the affected communities. The mean annual payroll at BRAC bases was just under $14 million prior to closure or realignment. The mean annual payroll after conversion was just under $3 million. The annual mean income plus the multiplier effect prior to BRAC action was $40 million. Following BRAC, it was just under $10 million.

Table 4.18 Difference of Means between Jobs Lost and Created

	Mean Difference	t-score (2-tailed test)
Jobs Lost from Base Closure	4865.2326	6.490**
Jobs Created from Base Conversion	896.0930	5.366**

N = 43
** = Significant at the .000 level.

Table 4.19 Comparison of Federal Drawdown and Local Planning (in months)

	Mean	Median	Mode	N
Length of Drawdown And Closure Process	42.2	36	36	32
Completed Strategic Plans	24	22.5	12	38

The mean length of the military drawdown as presented in table 4.19 above was 42.2 months, as opposed to 24 months for strategic planning. This indicates that local authorities were some eighteen months ahead of the military. The median sug-gests a difference of about fourteen months, while the mode indicates the amount of time between planning and withdrawal was two years in many cases. The sample size is also a telling piece of data. While a half dozen cases still had not completed strategic plans by January 2000, a full dozen were still experiencing military drawdown. Some of these had been ordered closed almost a decade before.

Clearly the military bears some responsibility for the slow pace of base redevelopment by not vacating the property soon enough. This probably reduces cost savings from closures as well. But what causes the slow pace of military drawdown? I believe planning for, and dealing with, environmental degradation probably has a great deal to do with this problem.

Table 4.20 Correlations of Environmental Problems and Closure Delays

	Drawdown Length	Degradation Delays	Pollution Index	Clean-up Index
Drawdown Length		.465**	.187	-.083
Degradation Delays	.465**		.402**	.276*
Pollution Index	.187	.402**		.419**
Clean-up Index	-.083	.276	.419**	

N=42
** = Pearson's r correlation is significant at the .01 probability level.
* = Pearson's r correlation is significant at the .05 probability level.

Table 4.21 Correlations between Pollution and Development Outcomes

	Jobs Created	Annual Payroll	Annual Payroll + Multiplier
Pollution Index	-.211*	-.132	-.267
N =	41	21	18

* = Kendall's tau c is significant at the .05 probability level (1-tailed test).

Table 4.20 demonstrates strong correlations between degradation delays, the pollution index, the clean-up index, and the length of the military drawdown. The pollution index is a compilation of the different types of contamination found at each former military base in the study. The clean-up index is a compilation of the different types of solutions employed to deal with the contamination problems at each site. The pollution index correlates strongly and significantly (Pearson's r = .402**) with degradation delays. Degradation delays correlate heavily (Pearson's r = .465**) with the length of drawdown. As we would expect, there is a negative relationship between the clean-up index and the length of drawdown, but it is not statistically significant. The clean-up index does correlate strongly with the pollution index (Pearson's r = .419**), suggesting that at least the local authorities have had some success matching environmental problems with corresponding solutions. Without this, perhaps the pollution-related delays would be worse than they typically are.

Table 4.22 Correlations of EIS, Grants, Drawdown, and Jobs Created

	Environmental Impact Statement	Amount Grant $ Received	Drawdown Length	Jobs Created
Environmental Impact Statement		-.189	.524**	-.188
Amount Grant $ Received	-.189		-.256	.520**
Drawdown Length	.524**	-.256		-.225
Jobs Created	-.188	.520**	-.225	

N = 42
** = Pearson's r significant at the .01 probability level.
*= Pearson's r significant at the .05 probability level.

To substantiate the notion that environmental variables are a good part of the reason why insufficient jobs have been created at former military bases, more analysis is in order. The pollution index was compared with the dependent variables of jobs created, annual payroll (since closure), and the estimated annual multiplier effect associated with payroll income (see table 4.21). The inverse relationships between pollution and the dependent variables are expected. The difference in the sample sizes is due to the unavailability of data regarding income and multiplier effects in many of the cases, which affects the statistical significance.

Despite the fact that pollution seems to be a pretty significant barrier for job creation at former military bases, there may be other procedural factors at work as well. Table 4.22 presents the negative effects of a lengthy environmental impact assessment, and compares it with the expected positive investment effects of seed money awarded through the grant-writing process.

As expected, the environmental impact statement process, measured in monthly increments, correlates heavily with the sluggish military drawdown problem (Pearson's r = .524**) suggesting that they may both be caused by similar factors, namely the assessment of the types and severity of contamination. A lengthy EIS process also negatively affects the amount of grant dollars received by the local development authorities (Pearson's r = -.189) and the number of jobs created (Pearson's r = -.188), probably because it delays and complicates strategic planning. However, these were not statistically significant. As expected, there is a strong, positive correlation between grant money and jobs created (Pearson's r = .520**).

Comparing Barriers and Incentives with Outcomes

It seems very clear that environmental problems can have harmful effects on the progress of base redevelopment, but it is unclear how pollution compares with other developmental variables. The above correlation tables indicate that there are negative effects, but that they may not be significant. Perhaps this is because they may only be temporary. But perhaps we need to look more closely at how the intervening variables interact with the causal variables. Ordinary Least Squares (OLS) regression analysis allows us such a comparison. The beta scores indicate the relative influence of each variable while holding the others constant, while the R and R-squared scores assess the accuracy of the overall model.

The following regression (OLS) analysis compares many of the relevant independent and intervening variables in the military base transition process and allows us to see which have the greatest positive, and negative, effects on the base conversion outcome of job creation. The index variables (pollution, intergovernmental barriers, and economic incentives) are compilations of several related dichotomous variables. The grant money received variable is in dollar amounts, and seems to crowd out some of the others, notably the strategic planning variable.

The remaining variables (environmental impact statement, hiring of the base environmental coordinator, acquisition of the former base property, completed strategic plans, and inclusion of the business community) are all measured in monthly increments. In other words, the longer it takes to complete any of these tasks, the more we expect to see a negative relationship. The length of drawdown variable was taken out because it tends to crowd out the environmental variables. The pollution index demonstrates the negative, statistically significant relationship we expected to find with regard to job creation (beta = -.380*). But, rapid hiring of a base environmental coordinator (BEC) can significantly counteract some of these effects (beta = .314*). This could also be said for rapid completion of an environmental impact statement, but it was not significant (beta = .318). The intergovernmental barriers index had the expected negative effect (beta = -.171) but was also not significant.

Failure to complete strategic plans in a timely manner will result in a negative relationship with the dependent variable (beta = -.323), but this was not significant. The greatest and most statistically significant findings were with regard to the grant money variable (beta = .668**), the economic incentives variable (beta = .585**),

Table 4.23 Regression of Barriers and Incentives for Economic Development

Dependent Variable	Independent Variables	Beta Scores
Jobs Created		
	Pollution Index	-.380*
	Environmental Impact Statement	.318
	Hiring of BEC	.314*
	Intergovernmental Barriers Index	-.171
	Acquisition of Former Base Property	.197
	Completed Strategic Plans	-.323
	Amount of Grant $ Received	.668**
	Economic Incentives Index	.585**
	Inclusion of Business Community	-.375*
Multiple R = .891		
R-squared = .793		
Adj. R-squared = .660**		

N = 44
** = Significant at the .01 probability level.
** = Significant at the .05 probability level.

and the length of time it took to get the business community involved with the conversion effort (beta = -.375*). Just like with the strategic planning variable, the negative relationship is expected because of the way the data are coded. Failure to include the business community in local developmental discourse is statistically as big a problem as pollution.

The adjusted R-squared of .668** indicates a surprisingly high score for the overall model. I suspect this is being driven by the ratio-level variance of the grant money variable. However, it should be kept in mind that the amount and timing of grant money is heavily dependent on federal evaluation of local discourse, organizational development, strategic planning, and environmental assessment. If grant dollars are taken as an indicator of legitimation, as opposed to just an economic shot in the arm, funding is a measure of implementation progress. The relationship between grant money and job creation can then be viewed as an indication of policy implementation success. Of course, this may only be in relative terms. As indicated in table 4.18, the number of jobs created through base conversion by January 2000 had replaced only about one-fifth of the total jobs lost through the BRAC process. But the expectation is that the number of jobs created at former military bases will increase as environmental and implementation barriers are removed.

Summary

Base conversion is a very time consuming, difficult process for local governments. Many of the difficulties associated with the implementation of reuse plans are caused by intergovernmental conflict. Despite its best efforts, the federal government causes much of the delay. Often these delays are a result of a lengthy military drawdown and environmental contamination, which according to this analysis are probably related in many cases. These problems have an overall effect on the number of new jobs created, subsequent income generation, and local economic multiplier effects.

Base conversion also represents the practical and theoretical fusion of discourse, complex implementation, and local control. While base closures are not the economic disasters they are often portrayed to be, it still takes a great deal of capital and organization to successfully convert a military installation to new uses. There are a number of procedures in which local governments must educate themselves, such as property acquisition, the organizational structure of base conversions, and environmental requirements. The inclusion of all concerned local, public institutions and individuals is essential for the success of transition efforts. In fact, inclusion represents discourse in practice, and is legitimized through the federal grant process.

There are three sequential steps to a successful base conversion. The first is the "dual track" retention and preliminary planning phase. The second is the strategic planning and organizational development phase. This phase is crucial, because if the planning effort is not inclusive, and if certain procedures are not carried out, the reuse effort will stagnate. The third phase is known as complex redevelopment, and involves the implementation of strategic reuse plans. Environmental difficulties can rear their ugly heads at any phase of this process. Generally speaking, the sooner the better, so they can be incorporated into the overall development plan.

One shortcoming of the statistical analysis in this chapter is that it does not really get at the severity of environmental contamination at the various former military bases. This will be demonstrated in greater detail by the case studies presented in chapter V. The statistical analysis does, however, allow us to see the causal chains in the process and between the different types of variables. Without this explication of the model, the cases are harder to understand. We now have a template that demonstrates how base conversion is supposed to work, and how Pentagon brownfields can be redeveloped for civilian use, despite past instances of contamination.

Notes

1. Although expertise is actually a top-down concept, specialized knowledge is necessary to avoid lengthy delays. As long as communication remains multidirectional, hiring outside experts does not corrupt developmental discourse.

2. The BTC may be out of a job at the end of his assignment, depending on his seniority and specialization.

3. Easton actually argues that values are allocated authoritatively, implying they must be distributed by government. However, Frey, as well as Dye and Zeigler (2000), argues that economic activity is controlled by elites and does not reflect the will of the people. Therefore, economics is conducted in an authoritative fashion without regard for democracy. Another way to put it is Knott and Miller's (1987) neoinstitutional definition of "who gets what with differing institutional arrangements," which intimates that those who make the rules determine the winners and losers in allocational politics.

4. Long-term strategic plans differ from the preliminary plans used to legitimate conversion efforts because they are much more detailed and focus more explicitly on potential barriers to implementation. Hence, environmental problems discovered during the EIS process and subsequent impact on goals and objectives are more likely to be discussed here.

5. The practice of Economic Development Conveyance (EDC) originated with the Pryor Amendment (1993) when it became apparent that local governments would not have the resources to pay fair market value for military property. According to former Senator David Pryor (D-Arkansas), GAO hands out a Pryor Award annually to those local officials who have engaged in the most creative redevelopment solutions at former military bases.

Chapter V

Pentagon Brownfields and Green Redevelopment

As mentioned previously, the federal government is an environmental polluter. This most often happens at military installations and Department of Energy sites (Jacob 1992). Of the 116 federal facilities on the EPA's National Priorities List (NPL) of most polluted sites in the United States during the early 1990s, 95 were military installations (Steppick 1994), most of them still operational. Part of the reason for this was because at one time the Pentagon viewed contamination as a necessary evil in the name of national security (Federal Facilities Policy Group 1995).[1] Perhaps it is also the case that the long-term health and environmental implications were not anticipated or considered to be very serious at the time (*Los Angeles Times* 2000). Whether it was based on negligence or ignorance, substantial pollution problems still remain from the time period extending from World War II through the end of the Cold War. Three dozen of the bases closed by the Base Realignment and Closure (BRAC) process were designated Superfund sites, with a third of those in California (*Los Angeles Times* 2000).

In accordance with current downsizing trends, the Pentagon has been reducing both its domestic and foreign base infrastructure. Bases in the United States are closed through the BRAC process without regard for past or present environmental contamination, since they all are required by federal law to be cleaned up anyway. This presents localities with considerable difficulties if their soon-to-be-closed installations are polluted. Developers and local governments do not wish to accept liability for problems they did not create.

The necessity of post-Cold War commercialization of former military bases requires substantially greater efforts at restoration. Typical contamination at military installations involves spilled fuel, oil, lubricants, hydraulic fluids, industrial solvents, pesticides, paints and thinners (Bergstrom AFBCA 1995). Other pollution sources can include heavy metals and water, nuclear materials and other improperly stored hazardous waste and munitions, as mentioned in chapter IV. Since the 1970s, the military has dealt with these problems under the Installation Restoration Program (IRP) of 1978 and the Defense Environmental Restoration Program (DERP) of 1984, among other statutory requirements, such as NEPA and the Endangered Species Act. However, according to the military, the DERP is not as effective as it could be due to underfunding by Congress (GAO 1998). Since the enabling BRAC legislation was passed in 1990, the military has attempted to coordinate IRP, DERP, and BRAC under the Fast-track Clean-up plan, so developmental and restoration goals can be realized (Federal Facilities Policy Group 1995).

This chapter returns to the military transition cases introduced earlier in the book. First, I discuss the methodology used in the case studies. Next, I present a section of four main case studies that involve site visits, discussions with local officials, or some other degree of participant observation. They are Austin and Lubbock, Texas, Jacksonville, Florida, and Fort Smith, Arkansas. These cases were selected largely for convenience, and to save taxpayer dollars with regard to travel. I have lived in or near three of these cities, Lubbock, Jacksonville, and Fort Smith, at some point during the course of this study. I made several special trips to the fourth, Austin, during its redevelopment period, prior to the construction of Austin-Bergstrom International Airport.

The third section examines five other cases of interest. Some are included for the sake of geographical representation, because they were Superfund sites during conversion, or both. They include Fort Worth, Texas, El Toro, California, Devens, Massachusetts, Portsmouth, New Hampshire, and Monterey, California. Fort Worth is a case that utilizes a combination of restoration strategies, including the NPL. I grew up near the El Toro Marine Corps Air Station in Orange County, California, having visited there when I was much younger. I subsequently followed the lack of discourse in the conversion effort for several years. The former Fort Ord in Monterey, California illustrates what can go wrong when a community is faced with extensive pollution. Devens, Massachusetts, and Portsmouth, New Hampshire, on the other hand, illustrate how degradation problems can be overcome by effective public discourse.

Methods and Data

Cases for this chapter include several different criteria, representation among the different branches of the armed forces, geography, variation in pollution levels, as well as differing political processes and policy outcomes. The research design allows for contrast between the cases without having "apples and oranges" or units of analysis problems with the comparison. To enhance analytical generalizability, the cases attempt to clarify several research propositions (see Yin 1989). The general research question this chapter attempts to get at is, What is the relationship between the different levels of government involved in environmental restoration at military bases? This question deals with both cause and effect, and has important implications for hastening the pace of environmental restoration, job creation, and overall sustainable development. It also has implications for redefining the relationship between levels of government as the top level devolves its responsibilities to the states and local communities.

A related proposition has to do with how states and localities can increase their institutional capacity to solve environmental problems. This may involve elements of contracting out to private providers, or it may simply involve state coordination of federal clean-up efforts with local base conversions. The best indication of whether intergovernmental remediation efforts are a viable alternative to the traditional federally mandated CERCLA (Superfund) process is to gauge the effectiveness of the environmental Fast-track Clean-up process as implemented by the BRAC clean-up teams. Although the statistics in chapter IV indicated that Superfund was effective at cleaning up former military bases, many communities do not want the delays of up to ten years associated with being put on the NPL. BRAC clean-up teams provide a speedy, state-centered alternative to Superfund, but they can both be used in conjunction with the other.

The main dependent variable, as in the quantitative analysis, is job creation. Each case discusses the number of jobs lost versus jobs created, based on the most recent data available from the General Accounting Office. The main causal variable is interlocal discourse. Without public discussions about the environmental problems, a community will not be able to complete strategic plans or environmental impact statements, and its redevelopment effort will be stopped dead in its tracks. The cases were selected based on environmental contamination, unlike in the quantitative analysis; both polluted and (some) nonpolluted bases were included. Contamination is the key intervening variable, and has considerable variation among the cases. Gary King, Ronald Keohane, and Sydney Verba (1994, 51-63) claim that qualitative methodology involves systematic observation, which can be used to compensate for a small sample size, such as the nine cases used in this chapter. Besides contamination, the empirical referents of discourse are derived from the case narratives for this type of systematic comparison with the policy outcome of job creation.

Another crucial intervening variable for this chapter is the nature of the clean-up solutions involved in the conversions, which is partially why the several cases are chosen from different states. As discussed in Chapter III, there is considerable variation in the capacity of states to assist localities in cleaning up pollution. For instance, the state of Texas's TNRCC (Texas Natural Resource Conservation Commission) played a crucial supervisory role in environmental reclamation activity at all the transition sites in its state. Other states were less engaged or less helpful, such as California, even though that state reputedly has higher pollution standards than the federal government.

The Fort Worth case is discussed for purposes of contrast between the intergovernmental and top-down implementation styles of environmental restoration. This case includes an interesting twist—part of the base property is a Superfund site while the rest is administered in a way that is similar to the other two Texas cases. This allows a comparison not only between cases, but within one as well. The fact of the matter is that the contingency view of environmental clean up dominates local military base redevelopment. States and communities will pick whatever solutions fit their situations the best, given the available resources, rather than subscribing to some notion of the "one best way" of implementing policy decisions.

Other variables involve different characteristics between the cases, such as the military functions of the bases along with their distinct pollution problems. The missions at NAS Cecil Field in Jacksonville, Carswell AFB in Fort Worth, Bergstrom AFB in Austin, Fort Devens in Devens, Massachusetts, Pease AFB in Portsmouth, New Hampshire, and Fort Ord in Monterey, California, were combat oriented. However, Reese AFB had only a single training mission. This explains why the contamination problems at the former Cecil, Carswell, and Bergstrom bases are much worse. Fort Chaffee, near Fort Smith, Arkansas, was a typical Army base, having a small complement during peacetime, but considerable space to expand during periods of conflict. The toxic externalities of multiple, combat-oriented missions are much harder on the environment than the solvents used to clean training aircraft, which was the problem at Reese AFB.

The problems of inadequate hazardous waste disposal, and the presence of fuels, oil, lubricants, solvents, thinners, grease spills, paint, and pesticides have contributed to the former Carswell AFB being placed on the Superfund's National Priority List. The nature of the conversion at the base is different as well. The greater Ft. Worth area was forced to deal with base closure, then its preliminary planning and organizational development efforts were superseded by federal intervention in an attempt to at least partially "un-BRAC" the facility by reestablishing an active military operation.

The city of Austin received full ownership of the former Bergstrom facility through property reversion, which allowed it to circumvent a number of constricting federal rules and regulations. The city of Lubbock attempted to gain title to the Reese property through a similar reversion, but was unsuccessful. Instead, an interlocal reuse authority, comprised of representatives from the city, county, and other local institutions, will have to negotiate with the Pentagon for an Economic Development Agency (EDA) grant before taking title to the property. This means that they have to deal with the alphabet soup of federal agencies and a number of prickly procedures, which may delay conversion plans.

Since the focus of remediation and conversion at the former Bergstrom and Reese AFBs is at the state and local levels, we will look at them together briefly. The city of Austin built an airport at Bergstrom, which opened in 1999, while the city of Lubbock has strategic plans for a multisolution industrial park, which have been implemented gradually over the long term. There are many differences between a single-solution and a multisolution transition. A single solution involves transforming the entire property into one large entity, such as an airport. A multisolution transition involves a more complex parcelization of the property into different zoning sectors. The strategic plan for Reese AFB involves recreational, commercial, and educational solutions for different sections of the former base property.

Levels of local environmental activism are different between the two cases. Austin is more politically active. This is true for environmentalism as well as many other issues, and is probably a result of Austin being the home of the state capitol and its central location in the state, which makes it a more cosmopolitan city. Lubbock, on the other hand, is more conservative politically and less activist. This may be a result of its location on the South Plains of West Texas, isolated from much of everything else in the state. This is not to say that Lubbock does not have environmental policy—as is confirmed by local groundwater policy studies (Somma 1993 and 1994; Urban 1992)—just that the focus is different. While national groups such as the Sierra Club and Earth First! were concerned about pollution at Bergstrom, only local groups were concerned about groundwater contamination at Reese. So, while state involvement in environmental remediation is a key constant between the two cases, there are many other differences between them.

The Jacksonville case is interesting for several reasons. Despite early political conflict between community leaders and citizens, a compromise was reached on redevelopment solutions. However, there were several pollution sites on the base that had to be dealt with, and to a certain extent, dictated part of the reuse plan. The conversion has involved the property being divided into separate jurisdictions, with the city of Jacksonville's Parks and Recreation Department, Jaxport, which is the local port authority, and neighboring Clay County overseeing various aspects of the transition. Aviation, industry, and recreation comprise most of the conversion activities. The Cecil Field Development Commission and Jacksonville's Economic Development Commission played supervisory and coordinating roles.

The cases of Fort Smith, Arkansas, and El Toro, California, have interesting parallels. Though the El Toro MCAS facility resembles the Cecil Field installation in Jacksonville in its former missions, infrastructure, and pollution problems, its interlocal redevelopment dispute more closely follows that of the greater Fort Smith area. The dispute over the El Toro redevelopment centered over whether it should be used for a new airport, which was favored by residents in north Orange County, or developed for commercial and residential uses, which was favored by citizens in south Orange County. The most logical venue for resolving this dispute—the county government—was besieged with the infamous Orange County bankruptcy scandal in 1995 and with problems associated with overdevelopment in the rest of the county. Finally, the Navy ordered a solution be crafted before they parceled the property and sold it to private developers.

Fort Smith found itself in a similar situation when it and the neighboring town of Barling were unable to come to a compromise on reuse solutions at the former Fort Chaffee. At issue was control of the property and a determination of which city would be saddled with the contaminated sites on the former Army base. After two annexation disputes and a lawsuit filed by the city of Barling against the city of Fort Smith and the base redevelopment authority, a compromise was reached between the parties. Just as in the El Toro case, higher levels of government were either uninterested or lacked the capacity to deal with the dispute between the parties. The current agreement, which was arrived at through mediation, enables each side to get most of what it was after. However, in both cases it took a decade for their disputes to be resolved, a sad commentary on what happens when antidiscursive forces are at work.

Data for this chapter were collected and triangulated through government documents, journalistic sources, and participant observation in the Texas Tech University Military Base Conversion and Community Assistance Program, the University of North Florida Military Adjustment Study Group, and the Municipal Solutions consulting group, of which I am a partner. It should be noted that this work is very contextual and has been ongoing since the spring of 1994, which explains why most of the variables are process-oriented. This allows a comparison of cases at the same points in the process, while minimizing temporal differences arising from BRAC decision dates and different drawdown and closure timelines. There are of course differences between the four site visit cases and the others, which is why they are dealt with separately. The three different environmental solutions—parcelization, BRAC clean-up teams, and Superfund—are represented in these case studies, often in conjunction with one another, as are most of the different types of reuse solutions, and typical barriers to redevelopment.

Former Air Force bases predominate among the cases studied, since the Air Force took the greatest hit in the BRAC process. Despite this, I made every effort to include at least one case from each of the other branches in this chapter. So while the Air Force is represented in the Austin, Lubbock, Fort Worth, and Portsmouth cases, the Navy/Marine Corps are discussed in the Jacksonville and El Toro cases, and the Army in the Fort Smith, Monterey, and Devens cases. Some of these cases have the added benefit of giving us a more balanced regional feel. It should be noted that most military bases in the United States are located in the sunbelt rather than other areas, making the cases in this study representative of the total population of bases, so I do not feel that any particular section of the country has been deliberately ignored. Because California was hit with a plurality of base closures I felt it was necessary to include at least one from each region of that state, hence the reason for the El Toro and Monterey cases, from southern and northern California, respectively.

Intergovernmental Implementation

The intergovernmental solution, known as the Fast-track Clean-up plan by base conversion practitioners, is how many localities deal with Pentagon-caused pollution. In most states, this process originates with each branch of the military's base conversion agencies, which appoint a Base Transition Coordinator (BTC) and a Base Environmental Coordinator (BEC) to oversee the preliminary conversion process at the installation, while the military completes the drawdown of its forces. However, from this point on, restoration is coordinated by the state environmental agency, with assistance from the BTC, BEC, and an EPA representative, which together comprise the BRAC clean-up team (BCT) for the Air Force, or the Installation Restoration Program for the Navy and Marine Corps. The Army Corps of Engineers supervises most restoration efforts at former Army bases, sometimes contracting out for services with private entities.

Concerned members of the community and environmental interests are typically appointed to a local Restoration Advisory Board (RAB), to assist with coordinating public opinion and feedback (EPA 1998). Variations in how each branch of the military organizes these basic structures are noted in the case studies where appropriate. In an example of regulatory devolution, the state environmental agencies enforce EPA and concomitant state regulations in order to reach compliance levels, and often assist with contracting out for restoration solutions when it is needed. Unlike CERCLA, which has fifteen steps, the Fast-track Clean-up plan has only two, which reduces the chances of error and time delays, while reducing the federal government's—and hence the taxpayers'—expense.

Austin, Texas

The former Bergstrom AFB provides the first, and best, example of how federal and state governments, in conjunction with private contractors, can facilitate environmental reclamation. The closure of Bergstrom was atypical because, compared to most base closures which are characterized by the mobilization of support in favor of keeping the base, the city of Austin requested the cessation of military operations. While the military believed that the missions at Bergstrom were still necessary, the community was in serious need of better air transportation. The existing airport had grown too small to serve the population of the greater Austin area, which rapidly grew to roughly a million people during the 1990s.

During World War II, Bergstrom Airfield was established as a SAC (Strategic Air Command) base. During the Cold War, the mission was redirected to the Air Force's Tactical Air Command (TAC). Besides combat missions involving the stationing of B-52 bombers, and F-4 and F-16 fighters, Bergstrom AFB was used for support, maintenance, and repair of military aircraft. There was a great deal of contamination associated with these efforts, including "the use of paints and paint thinners, adhesives, cleaners, lead-acid batteries, solvents and pesticides. In addition, diesel fuel, gasoline, and jet fuels were stored in underground and aboveground storage tanks" (EPA 1998, 2). While these pollution issues began to be addressed as early as 1982, the full extent of contamination was not known until after the base was designated for closure in 1991. This potentially could have had substantial adverse effects on Austin's transportation problem and the pace of local development.

Transportation related to overdevelopment has become a nagging issue in the greater Austin area over the last decade or more. While the local economy crashed during the 1980s, with the collapse of the oil industry and the savings and loan scandal, it began a recovery in the early 1990s. As the state capital, Austin is home to many government employees. It is adjacent to the University of Texas, which has close to 60,000 students at any given time and is one of the major research universities in the South. Sixth Street, which is frequented by college students and visitors from out of town, has become a mecca for entertainment. Austin has also become home to part of the emerging high-tech industry, and has not suffered as much from the bursting tech bubble as California's Silicon Valley, which is probably responsible for driving much of the new economic development and corresponding population growth. Dell Computers is a local success story, for instance. The suburbs of

Round Rock and Georgetown to the north of Austin have expanded considerably, as have the small, exclusive towns to the west near Lake Travis, such as Lakeway. Urban sprawl has affected the Austin area, as many workers live in these suburbs and commute to the city every day.

Unfortunately, the main artery for most traffic in the Austin area is I-35, which runs north to south, just east of the University of Texas, downtown, and capital areas. The old airport was located near this heavily congested area, contributing to the problem. The airport itself was also too small to deal with the increased demand for air travel by the early 1990s. Noise pollution from this facility was also becoming an issue (EPA 1998). Bergstrom AFB, however, was located further to the southeast and out of the way of the immediate congestion, while still remaining relatively accessible. This made it an attractive option to help alleviate the air-related transportation issues. For the long-term, converting Bergstrom to an airport was probably more of a necessity than just an attractive option. An added benefit was that additional development might be attracted around the new airport instead of in the already congested downtown and I-35 corridor sections of town, perhaps relieving some of the overcrowding.

Unlike in most communities, drumming up support in favor of closing the base was not terribly difficult in Austin. Politically, Austin is probably the most liberal big city in Texas, if not the entire South. Also, the local economy was not dependent on defense spending related to the base the way other cities are. With the university, the state capital, and the emerging high-tech industry, few would miss whatever economic effects the base provided, especially if the reuse plan provided for greater levels of growth at the facility and surrounding areas. An international airport would certainly seem to be the answer to several of the Austin area's economic and transportation questions.

When the second BRAC round came to pass in 1991, the city of Austin decided to make a case for including Bergstrom AFB on the closure list. While most closure- target communities were struggling to articulate strategies to retain their bases, Austin was offering its up on a silver platter, despite a potential economic loss of $400 million annually. What was wrong with these people? The Air Force initially did not wish to place Bergstrom on the preliminary closure list, but acceded along with the BRAC commissioners, since they would not have to fight yet another base retention effort. Air missions in particular are easy to redirect to another facility, which is what they did with the largely combat-oriented missions at Bergstrom. By closing Bergstrom, the Air Force would be able to keep other installations open and the city would get its airport at substantial cost savings, so it seemed to please everyone involved.

The city of Austin also had a special deal with the Air Force that would ease the pain of base conversion. Many federal regulations that contribute to the delays of other base conversions were circumvented by a reversionary clause in the original deed for the transfer of 2,892 acres between the city of Austin and the U.S. Army Air Corps in the early 1940s. Four additional parcels totaling 324 acres were purchased by the federal government from the city in four separate transactions between 1958 and 1975, (Bergstrom AFBCA 1995). This 3,216 acres comprised the totality of the Bergstrom property.

Unlike other communities that had assumed the military would never shut down nearby bases, the city of Austin had the foresight to include provisions in each of its contracts with the military so that the property would revert back to the city in the event of closure. Not only did the reversionary provision exempt the city from many of the inclusion and strategic planning approval requirements, but it also enabled an easy transfer of property without having to go through the Economic Development Conveyance (EDC) or Public Benefit Conveyance (PBC) processes. While Austin seemed to be on easy street compared to most base conversions, it ran into the proverbial brick wall when the environmental assessment was conducted during the military's drawdown, which was completed within a relatively short two years.

By 1993, when the drawdown was complete, and the extent of the pollution at Bergstrom was fully realized, the base was placed temporarily in the hands of the Air Force Base Conversion Agency (AFBCA) during the cleanup, since the military was liable for the extensive environmental damage. A BRAC clean-up team and an environmental Restoration Advisory Board (RAB) of local citizens and environmentalists were created to assess and remedy the contamination problems. Strategic plans to build an international airport on the site were delayed somewhat by the Federal Aviation Administration (FAA), which did not want to start construction on the airport until liability for the pollution had been established. The FAA provided a great deal of the $750 million needed to construct the airport. It reserved the right to withdraw funding with no advance notice if its rules and regulations were not followed to the letter. Therefore, the city of Austin took no chances on procedural questions during the cleanup and subsequent redevelopment.

During the conversion, both the Air Force and the community wished to see the environmental restoration completed posthaste. Luckily for all concerned, there were no radioactive materials on the base property. Most of the contamination was limited to chemicals in the form of fuel, oil, solvents, pesticides, paint, and thinners. However, these contaminants existed at a total of 481 different waste sites on the base, which were designated for cleanup. "Waste management units varied from a 16-acre landfill to trash dumpsters used as refuse collectors" (Bergstrom AFBCA 1995, 1). There is also the continuing danger of groundwater contamination. Bergstrom sits over the Colorado River Alluvium, which is an aquifer utilized for drinking water and general use. Increased population growth and economic development place more demands for clean water sources. Groundwater pollution must be cleaned by first pumping water to the surface, then filtering it through a treatment system, before it can be returned to the ground.

Despite the severity of these problems, Bergstrom was not designated as a Superfund site. But if there had been radioactive materials, it probably would not have been so lucky. Placement on the National Priorities List would have caused extensive delays. Though there were many contaminated sites, none was severe enough for Bergstrom to be placed near the top of the list, which would have meant that it would have had to wait a substantial period of time before its problems would have been addressed. With so many different contaminated sites scattered throughout the property, parcelization was extremely difficult, if not impossible.

The city and the local community were extremely grateful for being left off the NPL. However, the cleanup is extremely costly. Fifty million dollars was allocated for reclamation, $28 million for impact studies, and an additional $50 million was earmarked for long-term monitoring. The Air Force is responsible for the $128 million total cost (Bergstrom AFBCA 1995). These costs, when taken with the almost $800 million provided by the FAA to construct the airport, make the price tag for the Bergstrom-Austin International Airport close to $1 billion, virtually none of which was paid by the local community.

The state of Texas, through the TNRCC, was in charge of most of the cleanup and contracted out with the Center for Environmental Excellence (located at Brooks AFB in San Antonio) and Law Environment, which are both large, nationally recognized consulting and environmental restoration firms. Bergstrom is now certified as a clean facility according to the state of Texas. Air cargo operations were moved from the municipal airport to Bergstrom in 1997 and have increased at the rate of 20 to 30 percent annually (EPA February 1998). Though the environmental restoration took substantially longer than construction of the new airport, Bergstrom International Airport opened for passenger traffic only a few months behind schedule in 1999.

This new development presented the city of Austin with several other problems, however. For one, what would they do with the former downtown airport? While this airport remained open, albeit at lesser capacity, it was also used to house military reserve units. But rather than just having to worry about the conversion of one facility, which is what most localities have to deal with, Austin actually had to worry about two of them. Another problem was fulfilling the statutory requirements of the McKinney Act of 1993, which mandated a property screening process for the homeless and for addressing American Indian claims. While the property reversion generally exempted the City of Austin from such requirements, the politically liberal polity of the area demanded these types of problems be addressed. During the conversion, base housing was relocated to other parts of the city and made available to low-income families through low-interest loans. This also helped to address Austin's low-income housing shortage, keeping it out of trouble with the federal Department of Housing and Urban Development (HUD).

Noise pollution was also a recurring issue, which was one of the reasons constructing the Bergstrom Airport was attractive in the first place. While the city saved approximately $50 million in noise reduction costs by converting Bergstrom (EPA February 1998), there remained the issue of additional noise pollution as further development occurred near the new airport. To mediate this problem, the area was rezoned primarily for commercial development. This prevented construction of residential neighborhoods that would later mandate noise pollution controls to maintain their quality of life.

While most communities fight base closure, Austin was a rare example of a city with the foresight to understand the advantages of base conversion. Bergstrom-Austin International Airport is large enough to handle present and future transportation demands, and is close enough to the city proper, but far enough away not to contribute further to the local noise pollution issue. The Austin case is also a good illustration of how intergovernmental solutions can solve extensive environmental problems before they can derail economic development. While an estimated 927 civilian jobs were lost at Bergstrom as a result of base closure, 1,984 jobs were created at the new airport, a 214 percent recovery (GAO 2002). Thus the recovery ratio is a favorable 1:2.

Lubbock, Texas

Lubbock, Texas, is a medium-sized city of about 200,000 people. It is known as the "Hub City" because of its central location on the South Plains of West Texas. Because it is the largest city on the South Plains, smaller communities rely on it as a center of commerce and to provide certain services that they are unable to furnish for themselves. It has a diverse economy with health care, government, higher education, and agriculture representing the largest sources of employment and services.

The installation, located less than ten miles west of the city, was a pilot training base between 1941 and 1997. "Reese opened in June 1941 as Lubbock Army Airfield on 2,000 acres donated by the City of Lubbock" (Durbin 1993, 10). The original wartime mission included the training of fighter, bomber, and transport pilots. "The field was closed during postwar demobilization and reopened in 1949 as Reese AFB under the Air Training Command (ATC) located at Randolph AFB in San Antonio, where it became one of five Undergraduate Pilot Training (UPT) bases" (Durbin 1993, 11). At the time of the BRAC closure decision, the base was home to fighter and transport training and housed T-37, T-38, and T-1A aircraft. The UPT program graduated about 400 pilots annually. As a result of base closure, Reese AFB lost approximately 2,720 jobs, of which 1,650 were enlisted personnel, 500 were officers, and 570 were civilian employees (Durbin 1993, 12).

In the spring of 1994, few in the Lubbock area suspected that Reese AFB was in any danger of being closed. This complacency on the part of the citizens and political leaders may have been partly responsible for the base closure. Compared to the other bases that Reese was forced to compete against to remain open,[2] Lubbock was better able to absorb the loss of federal dollars, which made it a bigger closure target. Lubbock had escaped previous closure threats with minimal effort, which may have contributed to the relative lack of concern until it was too late. The other communities also had fought off previous closure threats, and were more concerned about the potential effects of closure, due to the fact that their local economies were more dependent on military spending. Since the Air Force includes possible economic effects as part of its evaluation, such as whether the town will be forced to shut down, Reese was in a comparatively more tenuous position than the bases in Enid, Oklahoma, or Del Rio, Texas.

While the city of Lubbock was aware of the possible closure threat, it took a low-key approach in the retention effort until the initial Air Force recommendation to close the base was released to the Base Realignment and Closure commission. Reese had survived previous closure rounds in 1988, 1991, and 1993, with little effort on the part of the city or community at large. There was little evidence to suggest that the 1995 round of closures would be much different. However, as more undergraduate pilot training bases were closed, beginning with Williams AFB in the Mesa/Chandler, Arizona, community, competition between the remaining bases became more intense.

When the danger of closing appeared imminent in early 1995, the city of Lubbock began its dual-track strategy of attempting to save the base while exploring reuse alternatives. The retention effort was highly publicized, with members of the city council and upper-level staff taking trips to Washington, D.C. to lobby elected officials and the Air Force. The city council also spent $200,000 on a consultant—defense lobbyist Christopher Lehman—whose job was to give advice and campaign full-time on behalf of base retention. The city did not publicly place any emphasis on lobbying the Air Education Training Command (AETC)[3] at Randolph AFB in San Antonio, Texas, which was the most important step in lobbying the military bureaucracy. However, according to sources in the city government, Lubbock contacted a newly retired AETC colonel, who spoke with the AETC on the city's behalf, apparently with little effect.

While the lobbying efforts were going on, it was rumored that conversion plans were being discussed within the city bureaucracy. Some of the suggestions that filtered out during this time were the establishment of an industrial park, a medical complex, and a retirement community. The early conversion effort consisted mostly of education about the base conversion process, with little emphasis on planning, specific solutions, or how to implement them. Conversion ideas would resurface with a greater degree of seriousness once the BRAC made its initial recommendations and final closure decisions.

By the spring of 1995, it was apparent that Reese AFB would not survive the budget-cutting ax of the BRAC. The unenviable position in which the community found itself during the closure process was that the disadvantages regarding the Reese AETC mission were easily quantifiable, whereas the advantages were more subjective and difficult to measure. Hence, the advantages were left out of the analysis. The way the criteria were weighted also may have had an impact on the closure decision. The economic impact of base closure was given a higher weight by the Air Force to prevent the wholesale economic destruction of small towns dependent on defense spending, which would not help Reese since it was located near a comparatively larger city than its competition.

Despite the efforts of the city to mobilize public support during the BRAC visit, it seemed that the retention plan would be ineffective. Lubbock has a sizable population of veterans and pro-military citizens that could have been mobilized to direct and participate in the dual-track retention and conversion phases. However, the retention effort had only limited grass-roots support. The failure of the effort to save the base placed a greater degree of importance on reuse planning. Following the closure decision, local public officials urged the community to pull together for the conversion effort. Although they advocated an inclusive planning strategy, politicians and institutions really were unsure about what to do next. Also, given the closed nature of the retention effort, they were somewhat farther behind than other communities in their preliminary inclusion effort.

The blame in Lubbock for the closure decision on Reese AFB was mostly leveled at BRAC, not the city government. Nevertheless, throughout the spring and summer of 1995, the city adopted a quietly introspective posture, maintaining a closed-door policy with regard to what to do next, despite public assurances to the contrary. It also began to isolate itself from other public institutions, and excluded the community at large from the early part of the conversion effort. This strategy did not help the city when it encountered unexpected problems.

In the weeks following the closure decision, the city government initiated contact with the Office of Economic Adjustment (OEA) to discuss the allocation of planning grant money. OEA reportedly awarded the city a small $31,000 grant to hire staff and work on conversion-related activities—but the money never arrived, and OEA contact Colonel Joe Gorman was reassigned and was not replaced by OEA in a timely manner. By maintaining an exclusive debate, the city sealed itself off from other sources of planning money, such as that which could have been raised by local financial institutions, private foundations, and professional business associations.

Part of the reason for the city's behavior can be attributed to the fact that they are the largest local governmental institution in the community, and as such it is usually responsible for providing leadership on most policy issues that affect the Lubbock area. Along with leadership comes the inevitable criticism when plans are not effectively implemented, or when externalities arise. It could be that the city was not attempting to behave in an exclusive manner, but that its introspective demeanor was designed to formulate strategies to ward off unwarranted criticism, and that this was misinterpreted by other local institutions. The city was probably shell-shocked from the unexpected loss of the base, while at the same time it was faced with having to formulate the best course of action as part of its dual-track strategy.

It should be noted that the city of Lubbock's track record on successful implementation of economic development or infrastructure-related ventures in the 1990s was not good. It had little success with the water reclamation project at Lake Alan Henry, located south of Lubbock, because the city neglected to negotiate the mineral rights to the land that was to be submerged by the lake. A number of oil wells remained on the property, and as long as they were functioning the dam was required to remain open. That dispute has since been resolved and the lake is open for recreation. Mayor David Langston's efforts to secure a Department of Defense accounting office and a local multipurpose arena in 1994 ended in failure, as the Pentagon awarded the accounting office to San Antonio and the multipurpose arena ballot initiative was defeated in an election. Plans to redevelop economically depressed parts of downtown and East Lubbock were also largely unsuccessful. Perhaps one reason for this consistent failure was the repeated exclusion of prominent local financial, real estate, and industrial interests from these efforts.

One way in which potential organizational problems could have been avoided would have been to expand the strategic planning committee and its subcommittees to include citizens not affiliated with political institutions, like other base conversions. Also, local business interests could have been directly included on the board. The officials involved counterargued that since they had largely come from the local small business community, they adequately represented financial interests. Although the city government did establish a temporary toll-free number for citizens to call with reuse suggestions, they made no effort to go out into the polity to draw in interested parties. Greater efforts at community outreach could have been attempted through the use of opinion surveys, the organization of veterans' groups, and racial minorities, and the establishment of regularly held "town meetings." A fax number or e-mail address might also have been helpful, but these types of activities were never publicly discussed or attempted at the time.

Token support from the business community was encouraged in late October by including twenty-nine members on the economic development subcommittee, but the appointments to this and the other subcommittees came comparatively late in the process and were made behind closed doors. Some of the subcommittees—although they were appointed in October 1995—did not meet until after January 1996. For the most part, there was not a great deal of attention given to the subcommittees. There was little media coverage, and citizens were not openly encouraged to attend and participate in their meetings. The practices of exclusion and organizational hierarchy indicate a lack of sincerity, possible hidden agendas, and little opportunity for willing engagement in the developmental discourse. While the participants may have been representative of the community, it does not necessarily follow that democracy was the result.

The difference of opinion over property reversion between the city and the Air Force was a significant development in breaking the exclusive city control over the base conversion effort. The original deed signed in 1949 by the city of Lubbock and the U.S. Air Force provided the Air Force with 1,387 acres, with the stipulation that the land return to the city in the event that operations at the base were terminated. In 1950, the deed was renegotiated. The reversionary clause was removed in exchange for a guarantee that the base would exist as a permanent installation. The city's position was that the original deed superseded the subsequent agreement, given the fact that the Air Force wished to terminate its guarantee of permanent occupation in the second agreement.

The Air Force believed that maintaining operations for forty-five years constituted upholding the intent of the second agreement. The city countered that "permanent" constituted some period of time longer than forty-five years, and pressed its claim to the property. After 1950, the base had expanded to include 3,954 acres, including an auxiliary airfield in Terry County, south of Lubbock County. There was also the issue of base housing, which was located directly across the street from the front gate of the base and was within city limits, as opposed to the rest of the property, which was on unincorporated Lubbock County land. Within a few short months, the city had given up its reversionary claim, forfeiting a potential shortcut to reuse implementation. However, it became much more open to inclusion after that.

The conversion of the former Reese AFB in Lubbock provides an excellent example of intergovernmental implementation of environmental restoration policy.

Like most communities, Lubbock did not wish to see Reese closed, nor did it wish to see it put on the NPL. Once the closure was seen as inevitable, the community, led by the city and county of Lubbock and Texas Tech University, attempted to construct strategic reuse plans and conversion organizations similar to the ones utilized by successful conversion efforts, such as those at Alexandria, Louisiana, Rantoul, Illinois, and Beeville, Texas. However, one possible impediment to redevelopment had been known for quite some time—the presence of TCE in local groundwater, under and around the installation.

Contaminants can easily flow from the surface into the groundwater aquifers on the high plains through playa lakes. These shallow lakes fill with water from thunderstorms, which often appear randomly throughout the spring and summer months during the monsoon season. The water then seeps into the Ogallala Aquifer, which runs from South Dakota to the Permian Basin in West Texas, and is the largest underground body of water in North America (Urban 1992). West Texans depend almost exclusively upon groundwater resources for consumption and agriculture. Groundwater contamination is the one environmental issue that concerns most West Texans (Somma and Hansen 1996).

Runoff from the washing of training aircraft carried cleaning solvents to the playas at Reese AFB. A plume of TCE was discovered in the groundwater near the tower on the east side of the base during the strategic planning phase of the conversion. It has since migrated under the base housing, which became mostly vacant after the 1995 closure decision. The plume forced the closing of a nearby nursery because the water was not fit to drink. This caused the nursery's liability insurance to be canceled, forcing it out of business. The most contaminated site on the base was located in the recreational area. The silt beneath Picnic Lake, which is really just another playa, was full of chemicals that needed to be removed. The site will have to be monitored for thirty to fifty years.

Lubbock County was happy to cede its interest in control of the property to the city when the severity of these problems was determined. While the county was not liable under federal law, County Judge Don McBeath was so uncertain of how to determine the difference between past and future contamination that he refused to take the possible risk. Seeing the revenue potential of the former facility, the city of Lubbock was willing to assume any risk. This worked out well for both institutions.

The Lubbock community had two strategies to deal with pollution problems. The first involved strategic planning. The second involves continued intergovernmental remediation. The Lubbock-Reese Redevelopment Authority (LRRA) accounted for the environmental problems in its strategic plan by favoring a multisolution reuse proposal. By rezoning the property for different solutions, they parceled contaminated areas out of the immediate development plans entirely. By separating these sites from the rest of the facilities, the redevelopment was able to continue with a minimum of delay from environmental problems or reclamation activities.

The BRAC clean-up plan was fully implemented at Reese. The TNRCC (the state's environmental protection agency) contracted with the Army Corps of Engineers stationed in Tulsa, Oklahoma, who subcontracted with Radian International, to filter the groundwater contamination. Radian is one of the largest environmental restoration companies in the world. It established a pump and treat system outside the northeast corner of the base housing to clean the groundwater plume. This may take as long as twenty years to complete at a cost of $10 million, not including long-term monitoring, according to the state of Texas.

With parcelization and the screening of potential new tenants to ensure that they will not additionally pollute the former installation, Lubbock was able to move ahead with its redevelopment plans, and should not have to worry too much about environmental liability concerns. However, it is debatable as to how much success their multisolution redevelopment plan has attained. Of the estimated 1,238 civilian jobs lost at Reese AFB as a result of closure in 1997, only 138 new jobs had been created as of October 2001 (GAO 2002). This is a ratio of roughly 9:1 jobs lost versus jobs created—a dismal 11.15 percent recovery rate (GAO 2002).

Jacksonville, Florida

Jacksonville is a city of slightly less than a million people located in the northeastern corner of Florida. Economically, it benefited a great deal from Cold War defense spending. The Navy historically had a sizable presence at three main installations for many years. The Mayport Naval Installation is a port facility for aircraft carriers, cruisers, and their support ships. Naval Air Station Jacksonville (NAS Jax) is an air and logistical station. The former Naval Air Station Cecil Field was home to several F/A-18 and S-3 squadrons. In addition, across the state line in southern Georgia is the King's Bay Naval Station, which is the home port of several submarine squadrons. The Army maintains a presence at Camp Blanding, which is to the south, between Jacksonville and Gainesville. The Florida Air National Guard has a small number of F-15s based at Jacksonville International Airport, which is located on the northern part of the city.

With the end of the Cold War, Jacksonville sought to diversify its economy through increased commerce. The local port authority, Jaxport, which is a municipal enterprise, ran the Blount Island port facility, Jacksonville International Airport (JIA), and two smaller airports, Craig and Herlong Fields until the air and seaport functions were separated. Jaxport is in part responsible for the city's rise as a regional commercial center. In addition, banking, finance, and telecommunications are fairly important sectors of the economy, particularly in the downtown area. Jacksonville is also a racially diverse city. There is a very large, politically active African American community located downtown and on the north side of town.

Jacksonville has had a consolidated city-county government with strong executive powers since the late 1960s. This was created to combat political corruption and to stem "white flight" from the inner city during the civil rights period, which was costing the city a great deal of tax revenue. The nineteen-member city council hears public concerns and legislates, but engages in little direct policy implementation. There is no city manager—the mayor and his appointed subordinates are the chief actors. The responsiveness of this structure may be a predetermining factor in the rush to encourage Jacksonville's economic development. While such an arrangement might be amenable to developmental interests, it can also have accountability problems, especially when trying to assess the causes and solutions to the problems of over-development and urban sprawl. Traffic, especially around the military bases, and water consumption during periods of drought or fire are major issues.

During the BRAC rounds in 1988 and 1991, the Navy put NAS Jax on the preliminary closure list because of the large number of civilian employees at the base, who typically earn higher salaries than active-duty military personnel, thus making them more expensive for the Navy. But on both occasions local politicians, reportedly including then-Congressman Charlie Bennett, negotiated with the base leadership and the naval command structure. As a result of these negotiations, they managed to get the facility off the list prior to its submission to BRAC, thus saving it from closure and protecting the thousands of civilian jobs located there. While they managed to save NAS Jax for the third time in 1993, the city was unable to rescue NAS Cecil Field from BRAC. The base was originally scheduled to close in 1999, giving the community six years to plan, but it did not totally cease all operations until 2000.

Eight thousand five hundred military and civilian jobs were lost due to the closure of Cecil Field. Of the three major installations in Jacksonville, Cecil Field probably has the least redevelopment potential, which added to the intractability of the transition problem. NAS Jax is located on prime real estate along the St. Johns River, which could easily be subsumed into the port infrastructure. Mayport sits right at the mouth of the St. Johns River on the coast, near the Florida-Georgia border. It could easily be converted to a cruise ship port, or integrated into the nearby Blount Island shipping facility. Cecil Field, on the other hand, being located on the Westside of town, prompted less obvious solutions. This is not to say that redevelopment was impossible, just more difficult.

Immediately after the closure decision, then-Mayor Ed Austin created the thirty-five-member Cecil Field Development Commission through an executive order. Its purpose was to serve as an agency through which communication with federal and state agencies would be maintained, and to coordinate the conversion plans as they were formulated, which is typical of reuse planning committees. However, its membership was restricted to primarily governmental representatives. Most were elected officials (city and county council members) from a regional, interlocal, five-county area. The commission also consisted of three state legislators, numerous appointed officials from local economic development agencies, and members of the Chamber of Commerce, representing local business interests. Commission meetings were held once a month for about two years (Cecil Field Development Commission 1996).

The Development Commission was not terribly large considering the region's population, especially when compared to some other cases, such as San Antonio, Texas, or Rantoul, Illinois, which involved around 150 people in their strategic planning activities. And it represented local elites on behalf of their corresponding institutions, rather than individuals working on behalf of the entire community.

Nevertheless, the commission still made an effort to solicit public opinion through their monthly meetings and newsletter distribution. A local interest group, the Citizens Planning Advisory Committee (CPAC) provided a vital public relations role for the commission. However, the structure of the planning effort's coalition was noticeably top-down. And although it was representative of a variety of institutions and local elite interests, its operating procedures are not what could be characterized as democratic or bottom-up.

While public opinion was solicited in this process, it was not directly institutionalized on the planning commission. This led to a rather vociferous dispute between the commission and local citizens. The controversy centered over whether to locate a prison on the former base property. Prisons are often seen as a quick fix for economic development problems, as they provide jobs and make politicians appear tough on crime. There are several instances where prisons have been constructed on former base facilities with great success, the former Chase Field NAS in Beeville, Texas, being a good example (Hansen 1996). However, they can also result in the NIMBY phenomenon and organized opposition, which is what happened in the Jacksonville case.

This dispute put the planning commission in an awkward position. Eventually it acceded to public opinion and canceled plans for the jail. A renewed attempt was made to attract the aerospace industry as the central part of the multisolution redevelopment effort, which is now (a decade later) beginning to result in some developmental success. There are many normative reasons why people might not want a jail located in their community. But perhaps the conflict could have been averted if the planning commission had directly involved a greater number of local citizens in the process, and downplayed institutional affiliations, as is recommended by the federal redevelopment agencies.

While the planning commission was strong on policy formulation, it was weaker on the politics of selling their solutions to the community and leading public opinion. City Council President Jim Overton, who chaired the planning commission, does not agree with this assessment. He essentially argues that the commission represented all relevant facets of the concerned community, and that through the built-in feedback mechanisms, the public interest was served. While this may be true, there was a large, organized group of citizens who did not feel that was the case. John Delaney, who succeeded Ed Austin and was mayor of Jacksonville during much of the base conversion, admits that from the beginning the jail was a bad idea, and the fight with local citizens was a learning experience for the politicians (Fritz 1998). Alberta Hipps, who was an organizer in opposition to the jail, eventually was elected to the city council based on the recognition and experience gained from this dispute, giving the council a more grass-roots orientation.

The community outreach part of the redevelopment effort improved greatly once the jail dispute was resolved. Coverage in the local newspaper and on television was regular and frequent. The *Florida Times-Union* allocated a staff writer expressly for coverage of defense-related matters in the North Florida/South Georgia region, including the Cecil Field redevelopment beat. The commission meetings, with the exception of five executive sessions, were open to the public, though only about ten people not on the board typically attended. The development commission also maintained a website and phone access for citizens. The executive director, Herb McCarthy, was always accessible to the public and the news media. In addition, during the planning phase, the commission distributed roughly 400 newsletters a month to citizens' groups, private businesses, government agencies, officials, and the general public to increase awareness and solicit feedback (Cecil Field Development Commission 1996).

The Cecil Field Development Commission received $6 million in grants from the federal government for seed money, which indicated federal legitimation of the reuse effort. Sections of the base property were allocated to local institutions through both a Public Benefit and an Economic Development Conveyance. However, the property was divided between the city of Jacksonville's Parks and Recreation Department, Jaxport (the port authority), and neighboring Clay County, which will oversee the wildlife preserve on the southern end of the property. This division of authority is related to the multifaceted reuse plan, which necessitates different types of expertise to be effectively administered.

The next big set of issues in the redevelopment of Cecil Field involved a myriad of environmental problems. There were nineteen different sites on the base with varying levels of pollution. Of particular concern was the former storage facility for nuclear weapons on the Yellow Water portion of the base. Other problems included landfills, motor pools, and sections that contained spent ordnance. Most of these sites were located near the main part of the base and the airfield on the south end of the property, which were to be key elements of the base conversion effort (U.S. Navy 1997).

The Yellow Water section, which is a very large part of the base located on the north part of the property, had originally been used as a gunnery range before its transformation into a top-secret nuclear storage facility. Getting a straight answer from the Navy about such activities is still difficult. "The historical storage of nuclear weapons could neither be confirmed nor denied, based on interviews [with Navy personnel assigned to the Yellow Water weapons compound]" (U.S. Navy 1997, 3-149). But enough private sources have substantiated local lore about nuclear weapons storage at that section of the base (Strickland 1997).

Throughout the Cold War there were stories about how nuclear bombs were loaded onto railcars for transport to the Mayport Naval Station during periods of high alert, to be deployed on aircraft carriers—a local variant of the "mobile Chernobyls" theme. While urban legends describing the transport of nuclear weapons through downtown Jacksonville and across the river in the dead of night are dramatic, they had no confirmation until recent years. Even the official NAS Cecil Field environmental impact statement authored by the Navy dances around the nuclear issue, despite being rather open about other environmental problems (U.S. Navy 1997).

The presence and preservation of cultural resource sites is also a concern for the redevelopment timeline at Cecil. There are sixteen archeological sites on the base, presumably the remnants of temporary settlements by the Timucuan Indians who used to inhabit the North Florida region. Half of these are located in forested areas near the main air station and should remain relatively undisturbed, unless there are major changes to the strategic reuse plan. There are two more such sensitive cultural resource sites located at Yellow Water near where proposed construction for recreational facilities is to take place. Four more potentially sensitive sites at Yellow Water are located near areas that have been proposed for future light industry, but should be safe for the time being. There are also several threatened or endangered species inhabiting unused parts of the base that should remain relatively protected (U.S. Navy 1997).

Types of contamination found at the former NAS Cecil Field included lead-based paint, asbestos, pesticides, medical and bio-hazardous wastes, oil-water separators related to maintenance and jet fueling activities, lead and copper drinking water contamination, polychlorinated biphenyls (PCBs), unexploded ordnance, and radioactive materials (U.S. Navy 1997). The base property is sufficiently large that the nineteen contaminated sites were parceled out relatively easily, without derailing redevelopment too much. However, future liability is a problem because it is not clear that future contamination resulting from aviation-related activities at the facility will be distinguishable from that of past Navy uses. This is common in base conversion cases, especially those involving aviation, where the original mission and reuse plans mirror one another very closely. Also, strategic redevelopment plans have taken pollution sites into account so as to negate them. Without including parcelization into the strategic planning process however, the reuse effort would stall until clean-up standards had been fully met.

To meet federal and state environmental standards, the Navy cleaned up sites that posed a direct threat to human health prior to closure. For the sites that were parceled out, however, the Installation Restoration Program (the Navy's version of Fast-track Clean-up teams) was employed with the assistance of the Florida Department of Environmental Protection (FDEP). The state and the EPA oversaw the removal of toxic and hazardous materials, capping of old wells and water lines, and the pump and treat process used to restore groundwater supplies. Since the state of Florida is known to have the toughest environmental standards among southern states, it should not be a surprise that FDEP was involved. However, the Pentagon's clean-up processes are designed to increase local institutional capacities by involving state environmental agencies.

What is perhaps more interesting and unusual in this case is the level of federal, not state or local, involvement. While the military typically identifies polluted sites and foots the bill for restoration, it generally steps back and allows local conversion efforts to implement clean-up solutions. Perhaps it is the case that at Cecil the Navy did not want to be held liable for long-term problems or take the blame for an excessively long conversion timeline. Or perhaps it was just easier for the Navy to clean up its mess during the drawdown process so as to cover its tracks regarding secret missions. In any regard, the level of cooperation between the Navy, EPA, and the FDEP, under the auspices of the Installation Restoration Program, is notable.

Strategic reuse plans call for a three-way division of the property based on previous function. The Yellow Water section to the north is in the process of being redeveloped for some light and heavy industry, commercial businesses—which are useful for their sales tax revenue potential—forestry management, an equestrian center, and other types of outdoor recreation, such as hiking and camping. Yellow Water is administered by the city of Jacksonville's Parks and Recreation Department, which flourished with new projects under Mayor Delaney's "Greening of Jacksonville" program in the 1990s. The main station to the south includes the airport and is administered by Jaxport. This part of the conversion is essential to job creation. Jaxport has plans to integrate Cecil into an intermodal transportation system that links air, rail, and sea transportation of commercial goods together. A CSX rail line runs through the property and downtown Jacksonville to the Blount Island port facility, making the endeavor a likely success. The third section, located in the southernmost portion of the property, was transferred to Clay County for use as a wildlife preserve. Whether it remains so for the long term will have to be seen.

Despite claims that the former NAS Cecil Field would become the "economic engine" of the city's Westside, job growth has been relatively slow. In fact, most of the city's new development has taken place on Jacksonville's Southside, near the beaches. Of the 8,500 total jobs lost, only 596 have been created—a ratio of roughly 14:1. However, when compared to only the number of civilian jobs lost—995—Jacksonville has a more respectable 60 percent recovery rate (GAO 2002, 47). Or to put it another way, this is a ratio of 1.7:1 estimated civilian jobs lost versus jobs created. According to the strategic plan, more than 3,200 new jobs are to be created over a twelve-year period (Cecil Field Development Commission 1996). If this goal is actually realized, the long-term recovery rate will probably be more effective than the realization of short-term objectives has been.

Fort Smith, Arkansas

The former Fort Chaffee was a victim of the 1995 BRAC round. The disposition of the property has been controversial, to say the least. This case is a perfect example of what can happen when neighboring cities refuse to engage in interlocal discourse. In addition to the territorial dispute, the contaminated sections of the base were "hot potatoes" that neither side wanted to touch. The former base property covers 71,273 acres located east of the city of Fort Smith, Arkansas, but only about 10 percent was developed and used by the Army. It was this section that was subject to BRAC action in 1995. The roughly 7,400 acres located between Fort Smith and the small town of Barling was attractive because of its redevelopment potential. As such, this section became the target of major litigation regarding two separate, contradictory state annexation laws.

Fort Smith has had a close relationship with the base since the 1940s, when it expanded to meet basic training demands for World War II. Fort Chaffee is famous for being the place where Elvis Presley was sworn into the Army when he was drafted in the 1950s, before being shipped out to serve in Germany. Hence, by association it became part of American popular culture. However, like most bases near the end of the Cold War, it began to downsize and the neighboring municipalities began to explore its redevelopment potential. A preemptive attempt at annexation by the town of Barling in the late 1980s touched off the feud between that city and Fort Smith, years before the base was ever slated for closure.

The nature of the annexation dispute revolved around the issue of developmental capacity. Fort Smith is the second-largest city in mostly rural Arkansas, with close to 100,000 residents. Barling is a small town of only a few thousand people. By most accounts, it does not have sufficient resources to redevelop a former military base. Most of the former base infrastructure is relic from the World War II era, and is in serious disrepair or otherwise in need of updating. The most pressing needs are new roads, water and sewer, and emergency services. Fort Smith has the resources to address these problems, whereas Barling does not. In addition, Fort Smith is located upstream on the Arkansas River, and hence is in a position to control Barling's water supply.

Prior to the closure decision, Barling annexed approximately 7,100 acres of Fort Chaffee. At the time it did not seem terribly relevant, but it became considerably more so after the 1995 BRAC decision. In 2000, Jo Carson, a state representative from Fort Smith authored a different annexation statute that maintained a city could not annex property it did not have the capacity to develop. As a result of the second annexation statute, Fort Smith was free to annex the former base property from Barling, because Barling could not show within a specified period of time that it could develop the land. This prompted the lawsuit *Barling v. Fort Chaffee Trust* (2001) to which the city of Fort Smith was also named as a defendant.

Because the two cities refused to cooperate with each other, they did not receive much seed money from the federal government for redevelopment. They were also unable to come together on redevelopment solutions. Fort Smith and Sebastian County thus pursued their own redevelopment strategy without Barling, and formed a redevelopment authority, known as the Fort Chaffee Public Trust. To its credit, the trust attempted to address some of these developmental issues, but fell victim to its own chaos.

The nation was exposed to the nastiness of Arkansas politics in the late 1990s, when the Hutchinson brothers—one in each house of Congress—led the impeachment battle against President Clinton over the Lewinsky affair. Interlocal politics in Arkansas are similarly divisive, and the problems of the Fort Chaffee PublicTrust became a further distraction from redevelopment. First, the trust's executive director. Phil Reeves was stabbed in the upper back by an unknown assailant while leaving the office. Later police discovered Reeves had been attacked by his girlfriend, not by political opponents or random assailants as he had intimated (Hughes September 1, 2001).

Next, Reeves sent an insulting memo to trust board members making unkind personal statements about Barling city officials and a member of the Sebastian County Quorum Court. According to reports, he blamed his lack of judgment on painkillers and high blood pressure medication he was on while recovering from his injuries. Though he publicly apologized, Reeves ultimately resigned as executive director as a result of these meltdowns (Hughes 2001). Not only were these developments embarrassing, but they also damaged the credibility of the trust and provided more distractions from what was already a difficult situation.

The federal government does not permit behavior like this to go on indefinitely, as it would like to see economic development happen at its former facilities. Communities have seven years to get organized if they want to assume control of the property under a no-cost Economic Development Conveyance. Hence, the greater Fort Smith community was under some pressure to resolve this dispute, or risk complete forfeiture of the land in question. Despite the lawsuit, the city of Fort Smith, the trust, and the city of Barling began to negotiate on a settlement. While there was an agreement in principal to divide the property, it was not clear how this would be done. This is where the environmental problems came into play.

The former installation contains forty-eight solid waste sites resulting from past operations and maintenance (Bennett 2002). Neither side in the dispute was excited about taking on these problems, but they needed to be addressed before property could be transferred. Most environmental restoration had taken place by 2001, but there were some problems still being dealt with in 2002, and some such as the TCE spill require long-term monitoring. Not only do problems such as lead-based paint, old plumbing and pipes, sewage lagoons, and municipal-type waste, all of which are present at Fort Chaffee, present pollution-related delays, but they also represent challenges to infrastructure development and maintenance.

While most of the buildings with lead-based paint were transferred to the trust, either for use or for destruction, there are more intractable problems still being dealt with. There are incinerator slag piles contaminated with harmful dioxins, a TCE spill, a pesticide spill, and municipal-type waste landfills to be cleaned up (Bennett 2002). Despite this, 92 percent of the 7,400 BRAC acres ordered for closure and transfer has met restoration requirements (Bennett 2002). The location of contaminated sites did play a role in the final settlement of the annexation dispute.

Mediation was able to solve the annexation dispute before a lengthy court trial could take place. The property was divided, with Barling getting the northernmost portion and Fort Smith taking the remaining east and west sections. The strategic plan calls for a multiple solution-type base conversion. Ultimately, plans call for an extension of I-49 to run through western Arkansas up to Kansas City. It would wrap around Fort Smith, through the Fort Chaffee property, and would be flanked by commercial areas. There will be light industrial, residential, and recreational areas on the former base (Fort Chaffee Redevelopment Authority 1998). Barling is to control roughly the northern third of this area and provide essential services.

The sewer system and lagoons, which are located on the northern part of the base, fell under Barling's jurisdiction. Barling is responsible for providing services to its section of the base, which may be difficult, if redevelopment does not occur rapidly enough to generate revenue. In addition, Barling must provide wastewater treatment for Fort Smith. Rates and charges were negotiated into the agreement so as to prevent future disputes. Fort Smith began providing emergency and other services for the trust shortly after the annexation of the property, and will continue to do so on its section (City of Fort Smith 2002).

While it may sound as though the property was divided and the two cities went their separate ways, in actuality the agreement between Fort Smith, Barling, and the trust forces them to cooperate. This agreement perhaps heralds the beginning of interlocal discourse between the concerned parties. However, it may have come too late to be of much benefit. On the other hand, it should be kept in mind that base conversion is a long-term economic development endeavor, so if in twenty years the strategic plan has been realized, then Fort Chaffee will be considered a success story.

With regard to job replacement and income generation, this case is short on results. The GAO study that has contributed such figures for the other cases leaves out Fort Chaffee altogether. However, it is safe to say that outside of staff hired by the trust and the Army Reserve station on the property, there has not been any job creation at Fort Chaffee. While there were environmental challenges in this case, and haggling over property lines took polluted sites into account, it was the lack of interlocal discourse that caused most of the delays.

Superfund Implementation

To briefly summarize the CERCLA process, for a locale to be declared a Superfund site, the level of contamination must be bad enough to warrant inclusion on the National Priorities List. An NPL site is usually one that has radioactive or extremely hazardous materials present, but this is not a necessary condition. Once a site has been placed on the NPL, it is dealt with on a "worst first" basis, which is sort of like a triage system. If a locality is placed near the middle or bottom of the list, it may take a considerable amount of time before its problems are dealt with. For communities attempting to replace lost jobs and revenue associated with base closures, such delays are unacceptable. In addition, the protracted CERCLA process preempts state and local efforts, while adding more regulations, which further slows the process. As base conversion is essentially a local-level activity, additional interference from the national government is generally unwelcome.

Fort Worth, Texas

The former Carswell AFB (currently Naval Air Station Fort Worth/Joint Reserve Training Base) is directly adjacent to a National Priorities List site. This example provides us with a useful contrast between the traditional top-down method of environmental policy implementation and the more contemporary intergovernmental approach. This case also entails state involvement, which is a key variable in some of the other cases as well, although, it should be noted that the involvement of the TNRCC in the Carswell cleanup was not as extensive as in the intergovernmental situations experienced in Lubbock and Austin.

When Carswell AFB was placed on the 1991 BRAC closure "hit list," the greater Fort Worth community began to plan for the task of base conversion, as most do when confronted with the inevitable withdrawal of locally stationed military missions. The mayor of Fort Worth established an eighty-member task force to assist and participate in this process. Included in this exercise were the city of Fort Worth, Tarrant County, and two smaller, nearby towns. However, the transition was interrupted during the next round of base closures, as the Dallas Naval Air Station was closed and its mission redirected to Carswell (Pine 1994).

During the preliminary planning phase a study was conducted by the Southern Methodist University (SMU) Law Review Association to assess the possible impact of environmental problems at the facility. It argued that one of the greatest barriers to swift, successful military base conversions is environmental contamination. Property cannot be transferred to local governments, private companies do not want to relocate to contaminated facilities, and citizens do not want to work or live near such sites. The SMU study was designed as a critique of federal Superfund procedures to determine liability for the expense and clean up of hazardous wastes, and to prepare for any unwarranted delays (Steppick 1994).

With the imminent return of the military following the closure of Dallas NAS, local conversion plans were largely scrapped. However, during this time a Lockheed Martin plant had already relocated to a parcel of the property and was allowed to stay. But this parcel was the most contaminated and wound up as a Superfund site. Although civilian redevelopment is hindered by environmental contamination, military uses are not—hence the quick return of military operations.

There is a plume of TCE located in the groundwater that flows from the neighboring Lockheed plant. The problem is being dealt with in two different ways. The first is through the CERCLA process at the Lockheed facility. The second is through the BRAC clean-up process, which is discussed in the next section. The TCE was released for many years from tanks and vats, and has oil and grit mixed in with it. Two thousand acres beneath the base and two landfills are polluted with these chemicals (EPA 2003).

The massive release of the pollution, rather than the toxicity, is what caused the Lockheed facility to be placed on the NPL, although it should be noted that TCE breaks down into a major cancer-causing agent if not treated within a few years, which contributed to the complexity and severity of the problem. The cost of remediation for the next ten years will come to more than $30 million. In addition, there exists the possibility for chemical migration, which is being dealt with by the EPA, the Air Force Base Conversion Agency, and the Navy at the Fort Worth NAS (EPA 2003).

One advantage of federalism is that it allows for policy experimentation at the state and local levels. The newly realigned Fort Worth NAS provides us with such a policy laboratory. The portion of the base that is not an NPL site is a third example of intergovernmental remediation at military bases in Texas. Besides the usual pump and treat methods to clean up the TCE and petroleum wastes, they are experimenting with bio-reclamation ideas.

One such experiment is the introduction of oil-eating bacteria (known as "bugs") and oxygen into contaminated underground areas. Another is the planting of cottonwood trees and native buffalo grasses, which naturally filter out harmful toxins from the water. Both types of vegetation have taproots as long as 100 feet, making them ideal to help clean the groundwater. Except for the problem of beavers occasionally eating the trees, these experiments were judged to have had an immediate, positive impact.

TNRCC was the lead agency in the BRAC clean-up effort, with assistance from the AFBCA, the Forest Service, and EPA. In addition to the supervisory role, the state has assisted with cost reduction and has attempted to coordinate with the EPA at the neighboring Lockheed Martin NPL site, to reduce mutual costs and duplication of effort. The experimental remediation efforts at the Fort Worth NAS will enable them to complete their task ahead of schedule, and ahead of the NPL portion of the facility.

Because of the contamination problems, and what was essentially a reversal of the 1991 BRAC decision by the 1993 BRAC commissioners, Fort Worth has not enjoyed as much success at base transition as it potentially could have. Although the facility is being used, the civilian jobs lost have been only partially replaced. Of the estimated 869 jobs lost only 541 have been created, roughly a 62 percent recovery (GAO 2002). This is a ratio of 1.6:1 jobs lost versus jobs created.

Monterey, California

The former Fort Ord is located on the Monterey Peninsula in northern California, just a short trip south from the San Francisco Bay area. The base was used primarily as a training and deployment facility for the Army's 7th Infantry Division. Its location on the West Coast made it useful for large troop deployments during the Vietnam War (Dardia et al. 1996a). It was also used for this purpose during the first Gulf War just before the closure decision was announced in 1991. Though it is home to one of the newest universities in the California State University system, a small airport, a research center, and two golf courses (DOD 1999), much of the rest of the property remains underdeveloped, due largely to environmental problems. Were it not for the practice of parcelization, it is doubtful there would even be this much development on the property.

The former base is one of twelve designated as Superfund sites in California, which suffered a total of twenty-nine closures and realignments from 1988 through 1995. Fort Ord is the largest of these installations, comprising a total of 28,000 acres. Of this, more than sixty percent of the land will be set aside as open space (*Los Angeles Times* 2000). As of December 2001, 7,200 acres had been turned over to the federal Bureau of Land Management (BLM) for recreational use (Associated Press 2001), because most of it was too polluted for other uses.

Contamination problems on the base include chemical wastes, polluted groundwater, and spent and unexploded ordnance. The severity of seventy-seven years' worth of spent or unused ammunition throughout sites on the property is what caused it to be put on the NPL. Similar to other cases in this study, the Army was subjected to litigation from local environmentalists, known collectively as the Fort Ord Toxic Project, who delayed redevelopment until the Army implemented tougher restoration plans (*Los Angeles Times* 2000). The price tag for restoration, which involves the removal of munitions, contaminated soil, and water filtration, has cost taxpayers $250 million, with long-term monitoring expected to take ten to fifteen years at a cost of $75 million (Associated Press 2001).

Two Rand Corporation studies in the mid-1990s documented not only the severity of the contamination, but also the detrimental economic effects on neighboring cities resulting from redevelopment delays (Dardia et al. 1996b; Rubenson and Anderson 1995). David Rubenson and John Anderson assert that were it not for the intervention of former Congressman Leon Panetta, who later become President Clinton's White House chief of staff, the section of the property that became Cal-State Monterey Bay probably would not have been transferred to the university system (1995). This would have put the Monterey case in even more dire straights than it already is. They make the point that the former Fort Ord is hardly the picture of base conversion success it is often portrayed to be (Rubenson and Anderson 1995). While Michael Dardia, et al. (1996b) make the case that most economic problems due to base closure are localized adjacent to the former installations, and that in most cases these areas recover quickly, the former Fort Ord has not shown this kind of resilience.

Though attempts were made to speed up restoration at a number of sites on the property, only limited progress was made, and state and federal regulators were often left in the dark by the military (Rubenson and Anderson 1995). The parcel of the base that became the university was certified as clean, but students still could not drink the water, and much of the rest of the property fell into decay (*Los Angeles Times* 2000).

For its part, the Army asserts that it has been hampered by the Endangered Species Act, which prohibits controlled burns in areas where spent, and possibly live, ammunition may be present. They argue that the burns are necessary to clear away dense brush so that removal can take place. The public is leery of such practices because controlled burns have raged out of control in the past (Associated Press 2001). This poses an interesting policy question: what is worth more, cleaning up the ammunition to facilitate commercial reuse, or preserving endangered species? Turning the land over to BLM may be the best compromise. However, failure to effectively clean up contamination has halted plans to expand the university and redevelop the remaining base infrastructure.

Despite the need for affordable housing in the area, much of the former base housing remains vacant and unused. Those who are hurt particularly by this are low-income workers who provide most of the labor for the local service industry. The housing that has been allocated to neighboring cities will likely be replaced by more expensive homes (Baum April 27, 2003). This, on top of the environmental problems, has put a strain on local developmental discourse. The roots of this dispute seem to stem from failure to include the local business community for more than a decade after the closure decision. Another problem seems to be the lack of capacity of the local small towns with regard to economic development planning and implementation. The housing crisis also ties into the contamination problems. For affordable housing to be built, old buildings with lead-based paint and asbestos will have to be removed at a cost of $70 million (Baum 2003).

Though the Pentagon attempts to put the best face on it (DOD 1999) in its descriptions of base conversion progress, the recovery at Fort Ord is much less than its potential would indicate. The estimated number of civilian jobs lost was 2,835, while the estimated number of new jobs created is 1,135. This is only a 40 percent recovery, or as expressed in a ratio, 2.5:1 jobs lost versus jobs created (GAO 2002). The lack of credible discourse involving citizen groups, environmental interests, and developers has undoubtedly contributed to the problem. However, the severity of contamination on the base and the fact that it is still a major Superfund site probably overshadow the discourse problems to a great extent.

El Toro, California

The dispute over the El Toro Marine Corps Air Station centered on the possible conversion to an international airport. The time between the closure decision and the actual closure date—from 1993 to 1999, a total of six years—gave Orange County the luxury of time and should have reduced the impact of delaying tactics employed by South Orange County cities. However, it did not. The dispute over what to do with the former air station involved quality of life issues such as noise pollution, economic and international trade, interlocal administrative conflict, and a regional north-south, old money versus new money struggle among county residents. Further complicating these issues was the Orange County bankruptcy scandal that erupted in late 1994, which called into question the administrative ability of the county, which maintained a decentralized, nonprofessional system dominated by elected officials. This system was blamed for the scandal, as well as for a lack of responsiveness and accountability.

In November 1994, county voters approved Measure A, an initiative that required a commercial airport be built on at least part of the former base property (Argyros 1995). The initiative just barely passed, with north county residents heavily supporting it, while south county residents opposed it in droves. Many Orange County leaders asserted that the county would soon outgrow the local John Wayne airport and that another air-travel outlet would be needed (Berg 1995). South county cities argued that a commercial airport would disrupt their idyllic quality of life (such as it exists) with additional noise and pollution, and that the county government was not competent enough to administer such an undertaking. A planning committee composed of the five county supervisors and thirteen others from the south county region fell apart due to this dispute. Because these issues were not resolved, the Pentagon was asked to step in and mediate the conflict (Milbourne and Pasco 1995).

In what was called a "Solomon-like decision" by the Pentagon, Orange County was granted the power to be the lead institution in the El Toro redevelopment effort. In the meantime, the federal government would fund two redevelopment studies—one for the county government acting under the auspices of Measure A, and one for the south county cities that opposed the airport plans. Rather than quelling the dispute, this decision seemed to fuel the fires even more, with both sides condemning the decision as soon as it was made (Milbourne and Pasco 1995). By July 1995, the county's reuse authority had been disbanded (Rubenson and Anderson 1995). The south county faction geared up for a more serious challenge to the county's authority on several fronts. Another ballot initiative overturned Measure A in an election held in the spring of 1996. Also, a number of alternative reuse suggestions were proposed, including a minimum-security federal prison, a Disneyland-like amusement park, and an industrial park similar to those at base conversions at Beeville, Texas, Rantoul, Illinois, and Alexandria, Louisiana. These cases are often considered prime examples of base transition success because of the large numbers of jobs created.

The south county-based opponents of the airport, known as the ETRPA (El Toro Reuse Planning Authority) made numerous charges at the institutions involved with pushing the airport solution. They charged both Orange County and the FAA with covering up the truth about the environmental impact of the proposed airport. They accused the county of excluding all but those who favored the airport from developmental discussions. ETRPA filed several lawsuits against the county, actually winning two of them. The most significant ruling proved that the county had in fact falsified information in the El Toro redevelopment's EIS. The county was ordered by the court to release 5,000 pages of documents, disclosing that 91,000 pounds of air pollutants would be generated each day from airport operations (Eckles 2000). The judge also awarded the ETRPA $425,000 to cover its legal costs.

ETRPA expanded its membership to include twenty-nine local institutions, including not just the original seven south county cities, but also school districts, chambers of commerce, homeowner and civic organizations (Eckles 2000). This was presumably done to provide a contrast in the levels of discourse facilitated by each of the two transition efforts. ETRPA also engaged in a public outreach campaign, sending out informative surveys, establishing a website, and setting up a toll-free phone line. However, when attempting to contact them through these means, my phone messages were not returned, and the website was nonfunctional.

The lack of developmental discourse did nothing to address the environmental problems at the base. The different interests fought over how to clean up a single plume of groundwater for six years and did not reach any kind of consensus. This pollution extends for three miles beyond the former base property to a housing development in neighboring Irvine (*Los Angeles Times* 2000). In addition to the groundwater plume, there are polluted soil source areas that continue to feed the plume. There were also at least eight other sites that were dealt with by the Navy/Marine Corps's Installation Restoration Program according to a RAND Corporation study (Rubenson and Anderson 1995). How to deal with these sites, besides the obvious need to minimize environmental damage and threats to human health and safety, was complicated by the lack of a strategic reuse plan for the former base. For instance, parcelization was difficult to implement because it was not clear under each plan how it should be done, or if it would even be feasible.

Institutional conflict erected barriers to discourse in the El Toro case. There was a lack of interlocal sincerity and trust between Orange County and the cities near the base, including Irvine, El Toro, and Lake Forest, as evidenced by the lack of any attempt at their inclusion. Local politicians regularly questioned the competence of the county to engage in complex implementation endeavors, due to the Orange County bankruptcy scandal in 1995 and its failure to adequately manage urban sprawl issues. There apparently were hidden agendas on both sides, and there were few substantive contributions, judging by the lack of policy outputs in the planning and implementation process. Also, Orange County, with its population of three million, may be too large to create a sense of "community," which is a necessary precursor to developmental discourse. Nevertheless, "town meetings" could have been held and broadcast on local cable TV stations. Public opinion surveys could have been taken on reuse alternatives. Planning and redevelopment organizations could have been constructed to include ordinary citizens, developmental interests, and politicians representing interlocal institutions.

The failure to establish discourse resulted in the failure to establish adequate reuse plans and delayed the base redevelopment. In addition, the longer the fight went on, the fewer federal redevelopment dollars were available for the reuse effort. When the south county faction triumphed by winning the Measure W election in March 2002, the proposed airport was effectively killed. Measure W rezoned the area so as not to include aviation. Immediately after, the Navy announced it was going to sell the property at market value (Pasco, Martelle, and Weikel 2002). The city of Irvine made plans to annex most of the 4,700-acre base for parks and "other limited development" excluding aviation (Pasco 2002; Oldham 2002). While El Toro was listed on the NPL, largely due to the groundwater plume originating at two aircraft hangars, EPA decided to let the military continue its clean-up efforts under the Installation Restoration Program (EPA 2003).

The El Toro case is a perfect example of how not to conduct a base conversion and illustrates the importance of interlocal discourse by the absence of its empirical referents of inclusion, self-regulation, and policy outcomes. Of the 979 estimated civilian jobs lost, only 252 had been created by the end of October 2001. This is a ratio of about 3.9:1, or a recovery rate of less than 26 percent (GAO 2002).

Devens, Massachusetts

The U.S. Army began operations on the eventual site of Fort Devens during the mobilization for World War I. Largely a combat training base, it was operational through all the major military conflicts during the twentieth century. The 1991 BRAC designated two sections of the base for closure and a third for realignment into a reserve unit (EPA 2001; Pike 2002). Operations at the base ceased in early 1996. Despite serving time on the NPL, Devens is considered a base conversion success story, illustrating that becoming a Superfund site does not necessarily mean death to redevelopment efforts.

Fort Devens was first listed on the NPL in 1989 (EPA 2001), two years prior to the BRAC decision. Hence, the interlocal community knew of the level of contamination before the planning and reuse authority was established. There were still concerns, however, because the base was located near four cities, two counties, and was only thirty-five miles away from Boston. While this location undoubtedly helped lure businesses to the former base after closure and realignment, contamination problems were potentially devastating.

According to the EPA (2001), the Devens facility contained 324 contamination sites. These included maintenance yards, landfills, polluted groundwater, gas stations, underground oil storage tanks, the former Fort Devens elementary school, and the airfield. Most of the pollution was in the form of chemical contaminants, such as PCB, or extractable petroleum hydrocarbons (EPH) resulting from oil and gas spills or leaks. Besides CERCLA activity, the Army utilized the Installation Restoration Program and the BRAC clean-up plan (EPA 2001). Part of the former base was parceled into a nearby federal wildlife preserve, while much of the rest was redeveloped for civilian use.

Why was Devens able to overcome these problems, when they might have stopped other conversion efforts in their tracks? The solution was to deal with both development and contamination simultaneously. Sustainable development, which balanced the need for job creation with the realities of environmental restoration, was the primary stated goal for the Devens redevelopment authority (Devens Enterprise Commission 2000). The planning and reuse authority embarked on "an extensive public participation process" (Devens Enterprise Commission 2000) to deal with these problems.

The community planning effort sought to establish a consensus on what sustainability meant to them and how it would be attained. Specific issues, such as energy consumption, water quality, employment, the nature of participation, and planning, were all seen as relevant issues for their base reuse process (Devens Enterprise Commission 2000). Furthermore, indicators were chosen for evaluation against nine different benchmarks to ensure that goals were being met.

Governor William Weld appointed the first redevelopment board for Devens in 1991. On the surface this would seem to be antidiscursive, but the board began an intense interlocal discourse soon after creation. Land use planning was undertaken by the Boston Society of Architects, along with public participation from four different reuse planning committees. Other groups such as Mass Development were added to help implement plans articulated through the public workshops (Devens Enterprise Commission 2000). By agreeing on the goals and having specific plans to implement them, the greater Devens community was able to surmount its environmental problems, and attain job creation sufficient to replace those civilian positions lost from base closure.

The state government is ostensibly the lead agency in the redevelopment process, probably because Massachusetts is a small enough state that much of its economy could have been adversely effected by prolonged recession in the areas adjacent to the former base property. The state government approved a multiple-solution reuse plan that enabled Mass Development to issue bonds and market the facility to a variety of different businesses, namely, high-tech, research and development, environmental corporations, industry, and recreation (Pike 2002). The Gillette Corporation is one of the largest employers at the former base. The federal government, through the Department of Labor, the Justice Department, and the Fish and Wildlife Service, has also made use of part of the facility for a jobs center, a prison hospital, and a wildlife refuge, respectively (Pike 2002).

During wartime, Fort Devens had anywhere from 7,000 to 15,000 active-duty troops, but these were mostly transitory. The civilian employment was 2,178 at the time of the closure and realignment decision in 1991. As a result of the conversion effort, 2,288 new jobs were created at Devens for a 105 percent recovery (GAO 2002). That is roughly one civilian job created for every one lost, with a few to spare.

Portsmouth, New Hampshire

The former Pease Air Force Base in southeastern New Hampshire was a victim of the 1988 BRAC, and hence was one of the first bases to close in the early 1990s. The property consists of 4,653 acres that straddle a densely populated corridor along the Atlantic coast, which in recent years has become something of a suburb of the greater Boston area. The former base is located within the jurisdictions of two towns, Portsmouth, which at the time had a population of 26,000, and the small town of Newington, which had a population of only 800. This situation was potentially difficult because of the competing interlocal interests. When rumor of the impending base closure first surfaced in the 1988 round, the two towns drew up separate reuse plans (Guskind 1990).

When the closure decision was announced in 1988, the state appointed the Pease Redevelopment Committee with local representation to handle the strategic planning efforts. This preempted the local reuse plans of the two towns, staving off potential inter-local conflict, but setting the stage for intergovernmental problems. While most base conversions are dealt with locally, regardless of the institutional capacities of the governments involved, smaller states are more likely to be directly affected by base closure economically, hence we would expect their governments to take more of an interest than would larger states. There are not very many examples of these small states to draw from, but New Hampshire illustrates this point rather well.

The planning committee hired the Bechtel Corporation of San Francisco to produce a master strategic plan of the property. Its plan contained some reuse options that were unpopular locally, such as a giant theme park, a regional shopping mall, an international trade port, and an industrial center (Maremont 1993). These plans were later scrapped in favor of demolishing base housing to make way for a commercial real estate development (NAID 1994). Soon, however, the planning committee gave way to the reuse authority.

The Pease Development Authority (PDA) was formalized by the state in 1992 to handle the base conversion, and received state authority to issue $50 million in revenue bonds, which the local towns would have been unable to do on their own. However, then-Governor Judd Gregg alienated the local interests by further politicizing the redevelopment process as part of his campaign for the U.S Senate. He appointed loyal cronies to top PDA positions, which angered local officials who urged that they hire development experts instead (Maremont 1993). Throughout the campaign, state officials predicted a decision by Deutsche Airbus to locate a $60-million, 400-job aircraft maintenance facility at Pease. Only after the election was over did they acknowledge the Deutsche deal was on hold (Maremont 1993).

The PDA was viewed with suspicion by the local governments, because it appeared to lack confidence in local abilities to bear the expense and organizational difficulties involved with base conversion. While most base conversions experience delays from the military or other federal agencies, as discussed in chapter IV, states can sometimes contribute to delays as well. The intergovernmental difficulties between the PDA, and the towns of Newington and Portsmouth, were perceived by local officials to be a primary cause of delay in this case. Using the base conversion as an issue in the Senate campaign probably also contributed to the appearance that there were hidden agendas involved on the part of the governor, whether that was actually the case or not.

The intergovernmental dispute was not confined to the state and local levels, however. The Air Force was pulled into the conflict when the environmental problems were discovered. The state and the Air Force both ignored the need for a transportation study to determine the impact of redevelopment on local air quality. In response to this perceived inaction, the Conservation Law Foundation, a local environmental interest group, and the town of Newington sued the Air Force to halt all development. During the land transfer, the Air Force also discovered twenty-three toxic waste sites on the base. A federal judge directed the Air Force to initiate a new environmental study. Rather than contribute to any further delays, however, the judge ordered that the property transfer continue in a timely manner. Parcelization allowed the isolation of the toxic waste sites so that the transfer could continue.

"The main hazardous wastes at Pease are solvents, oil and gasoline" (*Air Force Times* 1998, 14). The Air Force maintains that all wastes are contained at parceled sites on the property but that it may take fifteen to twenty years for total restoration to occur. State researchers and epidemiologists engaged in a public health study to help determine the overall impact on the public. Like Cecil Field NAS in Jacksonville, Pease AFB was home to a nuclear weapons storage facility. However, there seem to be no local health problems related to this practice, according to the study (*Air Force Times* 1998).

Despite the severe recession in the northeast during the 1990s, the New Hampshire coastal area was able to recover economically. Pease has since been converted to an industrial park, with Pan Am Airlines and eighty other businesses taking up residence as new tenants. The sections that were unable to be reused were parceled into a wildlife refuge (Associated Press 1999). Of the 400 civilian jobs lost to BRAC at Pease AFB, an estimated 4,367 jobs were created from base conversion, giving them a ratio of 1:10.9 jobs lost versus estimated jobs created, a recovery of more than 1,000 percent (GAO 2002). So, despite the intergovernmental disputes, the lawsuit, and pollution, Pease is an example of base conversion success.

Discussion

This chapter compared four main participant observation case studies with five other cases of interest. From this analysis we can draw several conclusions about the causal relationship between the variables involved with the restoration and redevelopment at former military bases. First, the presence or absence of situation-regarding public discourse is a relevant predictor of whether a community will attain its developmental goals. Discursive will formation can even overcome severe environmental contamination when the discussion focuses on how to address those problems, as demonstrated by the Austin, Devens, and Portsmouth cases, though these communities experienced varying levels of difficulty establishing developmental consensus.

The only case where the environmental problems seemed severe enough that discourse could not solve them was Monterey. Although, given the apparent failure to establish an inclusive, goal-oriented discussion process, this is difficult to say with certainty. The solutions chosen to deal with each set of contamination problems varied depending on severity, state institutional capacity, and the differing needs for specialization. The remainder of this section attempts to compare the different cases in a more systematic fashion for the sake of analytical generalizability (see King, Keohane, and Verba 1994; Yin 1989).

First, a discussion of the empirical referents of discourse is useful for our comparison. Next, barriers to implementation, which must be surmounted by effective public discourse, are discussed. Finally, a comparison of policy outcomes—environmental solutions and jobs created—will assist us in determining the effectiveness of base conversion policy.

The empirical referents (see Hansen 1998) attempt to get at the authenticity of discourse and can be used to compare cases analyzed both directly and indirectly based on certain indicators. As discussed in chapter I, these are inclusion, self-regulation, and policy outputs. Since many of these cases did not emphasize self-regulation (self-governance or self-determination), it seems apparent that discourse was elite-driven (see Hansen 2003). This is especially true in cases where the states were more heavily involved, typically the smaller ones, as demonstrated by the Portsmouth and Devens examples.

While elite control is not as ideally egalitarian as most of us would like, it does not preclude the fact that discourse did take place, even if it was mostly at a high level. This is perhaps an artifact of intergovernmental policy-making. Nevertheless, there is evidence of a more egalitarian inclusion, as indicated by the participation of local citizens and interests in the strategic planning phase and the welcoming of outsiders with redevelopment expertise, such as land-use planning or environmental restoration consultants, in many of the cases. These practices are often legitimized with federal grants. There are also clear policy outputs and impact, including environmental impact statements and job creation sufficient to replace those civilian jobs lost as a result of base closure or realignment (see table 5.1).

Table 5.1 Empirical Referent of Inclusion and Policy Outcomes

Case Study (Former Base)	Inclusion	Federal Grants	EIS	Jobs Created*
Austin (Bergstrom)	yes	yes	yes	yes
Lubbock (Reese)	yes	yes	yes	no
Jacksonville (Cecil Field)	yes	yes	yes	no
Fort Smith (Fort Chaffee)	yes	n/a	yes	no
Fort Worth (Carswell)	yes	yes	yes	no
Monterey (Fort Ord)	no	yes	yes	no
El Toro (El Toro)	no	yes	no	no
Devens (Fort Devens)	yes	yes	yes	yes
Portsmouth (Pease)	yes	yes	yes	yes

*Jobs created must be sufficient to replace those lost due to base closure.

As demonstrated in table 5.1, most of the cases in this chapter did involve some level of inclusion. Whether it was actual discourse is probably debatable in some instances, and not every case was required to be as open in its process as the others, such as the city of Austin, which had claim to Bergstrom AFB through property reversion. Others, through exclusion, managed to disrupt any semblance of discourse, which had detrimental effects on their conversions.

A good example of this is El Toro, where Orange County and the FAA were found by the court to have fabricated or deleted essential information in their EIS, which was thrown out. What we do know about El Toro's contamination problems comes largely from the EPA (EPA 2002).

Often the presence of litigation is an indication that inclusion has not taken place. But it is sometimes the case that malcontents or those who place institutional

Table 5.2 Barriers to Base Conversion

Case Study (Former Base)	Environmental Problems	Federal Delays	Interlocal Conflict	Lawsuits
Austin (Bergstrom)	yes	yes	no	no
Lubbock (Reese)	yes	no	yes	no
Jacksonville (Cecil Field)	yes	no	yes	no
Fort Smith (Fort Chaffee)	yes	no	yes	yes
Fort Worth (Carswell)	yes	yes	no	no
Monterey (Fort Ord)	yes	yes	yes	yes
El Toro (El Toro)	yes	yes	yes	yes
Devens (Fort Devens)	yes	no	no	no
Portsmouth (Pease)	yes	yes	yes	yes

interests above those of the greater community will seek to disrupt developmental discourse with lawsuits. This was the case with the city of Barling in the Fort Chaffee dispute. The greater Fort Smith, Arkansas community and the Fort Chaffee Public Trust had reached an agreement, but Barling held out in hopes of getting a greater share of the pie, delaying base conversion for a decade. This is an example of interlocal conflict, but interlocal disputes can arise and be resolved without litigation. Barriers such as these are indicated in table 5.2.

All the cases had environmental problems of some kind, which is why they were chosen for elaboration in this chapter. In most instances they were potentially damaging to the community's strategic redevelopment plans if they were not dealt with. As indicated in table 5.2, several of the cases experienced federal delays, often from the military's failure to clean up its pollution before vacating the premises, or from having the facility put on the NPL. In the Fort Worth case, a permanent delay to base reuse plans came from the return of the military.

Table 5.3 Environmental Restoration Solutions

	NPL	Parcelization	BRAC Cleanup
Austin (Bergstrom)	no	yes	yes
Lubbock (Reese)	no	yes	yes
Jacksonville (Cecil Field)	no	yes	yes
Fort Smith (Fort Chafee)	no	yes	yes
Fort Worth (Carswell)	yes	yes	yes
Monterey (Fort Ord)	yes	yes	no
El Toro (El Toro)	yes	yes	yes
Devens (Fort Devens)	yes	yes	yes
Portsmouth (Pease)	yes	yes	yes

While being put on the NPL can represent insurmountable delays, in the instances of Devens and Portsmouth, it did not represent long-term problems. The El Toro case was a failure, but while environmental issues contributed to this, being really did not.

Interestingly enough, all cases utilized the practice of parcelization, thereby excluding the polluted sections of their bases from the main redevelopment activities. This provides at least some anecdotal evidence to attest to the popularity of this practice. Another popular solution is military-related cleanup, either in the form of the Fast-track or BRAC cleanup teams, or the Installation Restoration Program. While under the IRP, it seems the military is more likely to do the job itself. Under both programs, intergovernmental solutions and contracting out are possibilities, depending on the nature of the problems and community needs.

Table 5.4 Job Loss and Economic Recovery

Case Study (Former Base)	Estimated Jobs Lost	Estimated Jobs Created	Percent Recovery	Ratio*
Austin (Bergstrom)	927	1,984	214.02	1:2
Lubbock (Reese)	1,238	138	11.15	9:1
Jacksonville (Cecil Field)	995	596	60	1.7:1
Fort Smith (Fort Chafee)	n/a	0	0	0
Fort Worth (Carswell)	869	541	62.26	1.6:1
Monterey (Fort Ord)	2,835	1,135	40.04	2.5:1
El Toro (El Toro)	979	252	25.74	3.9:1
Devens (Fort Devens)	2,178	2,288	105.05	1:1
Portsmouth (Pease)	400	4,376	1,091.75	1:10

* Ratio is of estimated civilian jobs lost versus estimated civilian jobs created.
Source: GAO 2002.

Table 5.4 allows a comparison of the dependent variable, job creation, between the nine case studies discussed in this chapter. The big winner is Portsmouth, New Hampshire, with a favorable job creation ratio of ten new civilian jobs for every one lost from base closure. This should come as no surprise, however, as Pease AFB was the first base to close as a result of the BRAC process, hence, they had a head start. Austin has done well creating two new jobs for every one lost at Bergstrom AFB. Devens, Massachusetts has broke even with a 1:1 ratio, but it has made a full recovery despite extensive pollution. The rest are somewhere in the job creation process. Some may never fully recover. El Toro and Monterey, California, most likely are the unfortunate losers. Lubbock, Texas, and Jacksonville, Florida, may attain their developmental goals, but perhaps not for some time.

Notes

1. The Federal Facilities Policy Group was an interagency cooperative endeavor cochaired by Alice Rivlin during her tenure as director of the Office of Management and Budget (OMB), and Kathleen McGinty, who chaired the Council on Environmental Quality. The group included representation from the Departments of Agriculture, Defense, Energy, HHS, Interior, Justice, EPA, NASA, and the Army Corps of Engineers, with input from President Clinton's Executive Office of the Presidency.

2. These included Del Rio and Corpus Christi, Texas, Enid, Oklahoma, and Columbus, Mississippi.

3. The Air Training Command (ATC) was changed to the Air Education Training Command (AETC) in 1992 as part of Pentagon reorganization. The headquarters remained at Randolph AFB in San Antonio.

Chapter VI

Conclusion

Both quantitative and qualitative analyses demonstrate the many potential pitfalls for redeveloping former military bases. All cases in both studies discussed in chapters IV and V were BRAC bases closed since the end of the Cold War. Cases that demonstrate both success and failure in creating new jobs and reusing former base property are useful to illustrate the importance of environmental variables at crucial stages of the development process. This chapter summarizes the theoretical perspective of the book and its findings, then provides a critique of the BRAC process itself for the next round of base closures, to take place in 2005.

Military base conversion is a positive sustainable development activity. Reusing former installations saves ecosystems by utilizing property that has already been developed for other purposes. Over the years, facilities that were on the outskirts of town, or even some distance away, have seen local communities grow to encompass the installations. Many bases have become prime real estate for commercial development. Rather than contribute to urban sprawl, closed military bases provide ready-made industrial parks, airports, universities, and other essential economic benefits, generally without additional habitat destruction. The BRAC process illustrates that entire facilities encompassing thousands of acres can be successfully recycled.

The redevelopment process also puts pressure on the military to clean up its pollution. While active-duty installations are also required to do this, closure hastens the restoration process, so as not to hamper the eventual property transfer to local communities. Many closure delays are a result of federal foot-dragging. Sometimes this is related to a lengthy military drawdown, but often it is due to difficulty in cleaning up the facilities so they can be safe for civilian business and residential use.

In many cases, base closure results in federal assistance to local governments so that they can carry out their primary function—facilitating job creation. Federal assistance often takes the form of redevelopment grants. But the Pentagon and other federal agencies have also provided a great deal of guidance in facilitating local discourse, strategic planning, budgeting, and long-term environmental monitoring, which are essential to developmental success.

Alternative Intergovernmentalism

Throughout this book I have attempted to make an argument regarding the efficiency and effectiveness of environmental policy in the United States. Previous patterns of environmental policy implementation were mandated from the federal government down to the states and localities, and are generally seen to be inefficient and expensive, according to much of the literature. Alice Rivlin's (1971) argument regarding the empowerment of states, localities, and markets provides a theoretical blueprint for alternative methods of environmental policy implementation. As the federal government attempts to downsize and empower state and local governments, it must also devolve its policy enforcement function, or else it risks becoming an impediment to state and local efforts at development. Lower levels of government already suffer from the ever-present problem of scarce resources, due in part to the failure of the federal government to address the states' Medicaid burden and their own revenue shortfalls.

The key to addressing this dilemma is to promote intergovernmental solutions that facilitate certain types of policy devolution. While this was a trend in the 1990s with certain regulatory policies, large-scale policy devolution in the United States did not occur because the role of each level of government was not well defined in the policy implementation process. Nevertheless, the concept still has merit if and when it is carried out effectively. However, devolution must be done with care. For this to happen, the federal government needs to be more willing to share its revenue and its power with subnational governments.

One of the advantages of federalism is that lower levels of government can be used as policy laboratories. This feature of government is underutilized. The problem is that there are those who do not want to give up power centralized at the national level on the one hand, and those who would simply decentralize or privatize important functions, without crucial financial or administrative support and oversight, on the other. This could create more unfunded mandates that states and localities cannot afford, or set up a situation that unfairly benefits special interests, which is contrary to the public interest entirely.

The federal government has the resources to adequately fund and assume the risks associated with base conversion and environmental restoration. States have the capacity to coordinate and facilitate the efforts of agencies and contractors at different levels of government. Localities have the need and the desire to expand present economic development activity. They also have the ability to work closely with states and are fully capable of contracting out with other government agencies or private firms to increase their institutional capacities.

The Fast-track, or BRAC, clean-up plan is presented to demonstrate the advantages of such intergovernmental policy-making. According to the General Accounting Office, this method has saved the federal government "150 years of clean-up project work and $150 million in costs" (GAO 1997). Under the Fast-track Clean-up plan, the three levels of government must coordinate for it to work. State environmental agencies lead the clean-up effort, but must work with local officials who are in charge of economic development. Because of its liability, the federal government picks up the tab. Each level does what it does best in this process. The federal government covers the expense, the states regulate, and localities create jobs. Administrative discourse between the levels of government is essential to making this process work, and is legitimized by federal development agencies.

State Variations

There are potential pitfalls to devolving environmental and subsequent developmental policies. As Evan Ringquist (1993) and James Lester (1995) indicate, there is considerable variation in the capacities of states to carry out environmental policies, due to such factors as wealth, political culture, agency capture, and the like. However, as the findings of this study indicate, institutional capacity is only part of the picture. Discursive will formation also matters, as does the severity of the pollution. States that traditionally have weak environmental records, such as Texas, have done well at former military bases, whereas states with stronger capacities, such as California, have not. However, Massachusetts and New Hampshire did quite well, as we would expect.

For Texas communities, the intergovernmental BRAC clean-up process has worked well to facilitate environmental restoration and economic development at a number of former military installations. The Texas Natural Resource Conservation Commission (TNRCC) tried to find the quickest way to clean up environmental pollution at contaminated sites. It also made sure that a thorough job was done the first time so that they did not have to return for additional cleanup. This was done because the state of Texas and the local communities with which it coordinates its efforts simply do not have the resources to enforce and monitor inefficient, expensive clean-up efforts.

The two-step process involving the creation of an intergovernmental team and contracting out provides fewer opportunities for costly mistakes to be made and encourages experimentation, as opposed to the top-down, fifteen step CERCLA process, which is notoriously inefficient and expensive, according to the literature. The cases of Lubbock, Ft. Worth, and Austin contrast well with that of the NPL site at the Lockheed plant, to demonstrate the differences between the two styles of implementation. Although the problems with the Superfund are well established (Davis 1993; Hird 1994), there are few plausible solutions being advocated. Using BRAC clean-up teams or something like them to address brownfields redevelopment is something that deserves greater exploration.

California is a state that typically leads the rest of the country in increasing environmental standards. However, in the Fort Ord and El Toro cases, state involvement was notably ineffective. Perhaps in the case of Fort Ord, pollution was so severe that there was little that could be done. But the reality seems to indicate that there was a notable lack of authentic discourse at all levels of that particular conversion effort. Were it not for the intervention of Leon Panetta, the university, built on the least contaminated parts of the former facility, probably would not exist (Dardia, et al. 1996b). Because of the university, Fort Ord is often mentioned as a success, but given the environmental problems and the lack of any other development, this claim is misleading.

The El Toro case illustrates what can happen when there is no consensus at all about what to do next. Environmental problems, both real and potential, were used to kill the proposed airport. This leaves Southern California with future infrastructure problems and current pollution, both of which hamper development. In both cases, the state did not really provide much assistance to resolve these dilemmas. Perhaps the state of California simply had too many base conversions to deal with at once. It may also be the case that California's economy and budget crisis at the time did not allow it the luxury of doing so. Many states may face similar fiscal dilemmas during the next round of base closures.

Massachusetts and New Hampshire did well in attempting to reconcile developmental discourse with environmental restoration. In fact, the base reuse effort at Devens, Massachusetts, made sustainable development the prime goal of base conversion, according to its own literature (Devens Enterprise Commission 2002). They even set up a systematic scheme for measuring, defining, and evaluating whether they had attained sustainability. In both the Pease and Devens cases, the state governments played pivotal roles in assisting local discursive will formation and in helping to clean up pollution in order for redevelopment to proceed in a timely manner. This is probably due in part to the fact that these are smaller states that would feel the economic impact faster than large states such as Texas and California. Consequently, they became active in the process much sooner.

With regard to greening Pentagon brownfields, my policy recommendations are not unlike those of Ringquist (1993), who argues in favor of greater empowerment at the state level to address environmental concerns. However, my argument involves a fundamental redefinition of the relationship between levels of government and patterns of implementation within the U.S. federal system, and highlights innovative ways in which states and localities can enhance both their developmental and environmental capacities. Key to this is the establishment of situation-regarding discourse of the type advocated by Charles Fox and Hugh Miller (1995). Only then can there be sustainable development of the kind Shutkin (2000) refers to in his theory of civic environmentalism. Perhaps more important, the hard lessons learned from both success and failure at base conversion can be applied to other types of brownfields restoration and development.

Environmental Retrenchment

Despite progress related to intergovernmental policy implementation in the 1990s, the current trend seems to be in the other direction. Rather than devolving authority to states and localities to deal with military-related pollution, the current Bush administration is bent on gutting the very statutes that enable environmental restoration at military bases. This is being done in the twin names of national security and the war on terror. While this plan largely effects active-duty bases, some of these installations will undoubtedly be closed during the 2005 BRAC. Given that this will probably stall redevelopment during conversion, it would seem for a number of reasons to be a rather shortsighted notion. While we seem to have found solutions for reconciling development with the environment, we have yet to find a balance between the military and nature, which is what causes the Pentagon brownfields problem in the first place.

RCRA

The Resource Conservation and Recovery Act (RCRA) of 1976 was designed to clean up pollution on active-duty military installations. Key to the enforcement of this act is a provision that allows states to litigate when the Pentagon and the EPA fail to comply. This is not characteristic of other environmental laws effecting the military, which rely almost totally on EPA enforcement. The Army has requested and received support from the Bush administration and the EPA to repeal the section of RCRA that includes explosives, munitions, and shrapnel as toxic waste (Shogren 2003).

The rationale for these changes is the claim that such policies interfere with military readiness. States, such as Colorado, which filed suit against the Army in 1997 for leaving live ammunition on the site of the former Lowry bombing range, are concerned for the health and well-being of nearby residents and development (Shogren 2003). However, this can only mean negative publicity for the Army and any other branches of the military that refuse to comply with RCRA requirements. David Sorenson (1998) makes a strong case that the military generally wishes to remain a good neighbor with local communities. This would seem to apply to environmental questions, as well as to a number of other issues that Sorenson raises, such as disorderly conduct.

Due to liability concerns, failure to enforce environmental statutes, whether in the name of readiness or not, may cause reluctance among communities to redevelop former installations after the next round of closures. While the Pentagon has assured Congress that nonenforcement of RCRA will not effect clean-up efforts at bases that have been closed or those that will close in the future, the state attorneys general have substantial doubts (Shogren 2003). If the federal government refuses to comply with the current law, this could create difficulties for state-coordinated BRAC clean-up teams (BCTs), and lengthen the already cumbersome military drawdown process for facilities closed or realigned in the

2005 BRAC. Such outcomes will probably contribute to higher restoration costs, a greater likelihood of military sites being placed on the NPL, and longer base conversion timelines, reducing the pace of reuse and slowing rates of job creation.

Endangered Species

In May 2003, each house of Congress passed contradictory versions of bills that would exempt military installations from sections of the Endangered Species Act of 1973 and the Marine Mammal Protection Act of 1972. While the Senate version attempts to reconcile the need for readiness with environmental protection (SA722 2003), the House version would totally remove the military from accountability with regard to endangered or threatened species on military lands (HR1235 2003). Complicit in these actions would be the defense secretary and the secretary of the interior. The proposed legislation would also exempt military installations from EPA studies on the status of threatened or endangered species unless the studies had the prior approval of the Interior and Defense Departments (HR1835 2003).

While it may be necessary in certain instances to waive sections of environmental statutes for the sake of readiness, it is disturbing to see an attempt to give the military a blanket exemption at the hundreds of facilities still in operation around the country. One instance in which an exemption would probably be justified would be the Camp Pendleton Marine Corps training base near San Diego, California. At this facility, the Marines are restricted from practicing beach landings because of the threat to endangered species and habitats located there. Since this is one of the few bases where Marines engage in basic training and amphibious landings, it would make sense to grant them an exemption for the sake of readiness. However, there are likely other cases where the military just simply does not wish to comply with the law, or does not care to seek an exemption to the ESA.

With regard to the Endangered Species Act, states have little means to ensure that the military complies. While military readiness is important for protecting the country, it seems cynical and probably unnecessary to give the Pentagon a blanket waiver from environmental protection requirements. Surely there is room for compromise between the two positions, as the Senate bill demonstrates. Exemptions to environmental protections can be given to the military when they are deemed necessary, but to start excluding entire agencies or executive departments from the law sets a negative precedent. What is perhaps the most disturbing is the lack of accountability inherent in the military's requests. EPA will seemingly turn a blind eye to law enforcement, while the Interior Department can only check up on the military with the defense secretary's permission.

The Future of the Base Closing Process

Over a decade ago, Steve Koven (1992) wrote about how the Base Realignment and Closure process attempted to reestablish the Wilsonian pattern of separating partisan politics from administration through the establishment of independent commissions. Koven's analysis of government documents that outlined the process and its variations between each branch of the U.S. military found that despite appearances, base closure is a very political undertaking from beginning to end. But this is by no means the only word on the subject. Ken Mayer followed this up with two articles (1995, and 1996 with Anne Khademian) that demonstrated that while independent commissions such as BRAC displace politics away from Congress, defense policy is still much more political than previously acknowledged in the literature. David Sorenson (1998) considers both of these views, but comes down on the side of the policy learning argument, making the case that the process has changed continuously over time to aid communities dealing with base transitions.

Congress, after much wrangling with downsizing proponents in the Bush administration, has grudgingly accepted another round of base closures but has postponed them until 2005. The remainder of this chapter deals with the question of whether there should be more base closings and how they should be conducted. This chapter concludes with some policy recommendations to streamline the closure and realignment process.

Base Closure Rationale and the Fight for Retention

Why should the U.S. government close more bases? And why do representatives fight closure decisions? The answers to these questions are complex, as are the politics involved with the closure, realignment and conversion processes.

Military bases are known to bring numerous jobs and economic benefits to neighboring communities, especially those bases that employ a great number of civilians. These include maintenance depots and logistical (especially Air Logistics Centers) facilities. A large domestic military base infrastructure has also been seen as essential to national security since World War II, when many of the current installations were constructed. Though many of these World War II-era bases were decommissioned between 1946 and the beginning of the Korean War in 1950, most of them were reopened for the duration of the Cold War. Though the Pentagon did close bases during and just after the Vietnam War, most of the domestic base infrastructure was seen as politically untouchable by the 1980s.

Since 1988, when the Cold War began to wind down, the Defense Department has argued with Congress over the need to maintain such a large and expensive domestic installation capacity. The Pentagon believes it is bloated with an excess of facilities that are a drag on the current war effort and future missions, and that it should downsize by 30 percent. The four previous BRAC rounds only reduced capacity by a total of 15 percent, in contrast with the 30 percent reduction in active-duty personnel (DOD 2001a). DOD is especially concerned with overcapacity at those bases that employ large numbers of civilians. Privatizing the work done at domestic military bases potentially increases operating costs rather than reducing them, presumably because salaries and benefits for civilians are more expensive than they are for active-duty military personnel. When faced with a competitive bidding process, government service providers actually deliver services cheaper than private contractors (GAO 1999a and b). However, pork barrel politics requires additional spending that neither the voters nor the Pentagon actually want. As a result of such issues and pressures associated with providing new missions on top of old ones, the DOD would like to see a reduction of an additional 15 to 20 percent in their institutional capacity (DOD 2001a).

Congress, on the other hand, has opposed Pentagon requests for additional downsizing. Though no congressman has ever lost an election as a result of base closure (Sorenson 1998), they fight it just the same. The reasons for their opposition to Pentagon downsizing are generally couched in economic terms, but they may simply not want to appear unpatriotic, especially during a climate of pro-military fervor. However, Defense Secretary Donald Rumsfeld and others in the George W. Bush administration insist that Congress's failure to pass additional closure rounds in 1999 and 2001 is actually a drag on the current war effort and future strategic abilities (DOD 2001a). Hence, they proposed more base closures to take place in 2003. Despite the best efforts of Representative Vic Snyder (D-AR) and Senator John McCain (R-AZ), who proposed two additional closure rounds for 2003 and 2005, the soonest Congress would agree to more closures was 2005. So disgusted at the amount of pork barrel spending in the final conference committee version of the 2003 Defense Bill—which contained the base closure amendment to the 1990 BRAC legislation—John McCain voted against it (Skornek 2001). With typical bombast, he proclaimed, "[t]he Appropriations Committee is intent on using the Department of Defense as an agency for dispensing corporate welfare" (Curry 2002). I tend to agree.

The most recent amendments to the original BRAC legislation contain several new provisions designed to reduce the need for communities to hire consultants for their base retention efforts. What this does is set up a pre-BRAC step designed to protect certain bases, roughly half the total number, from scrutiny during the closure decision process. Congressman Snyder in particular would like to protect those bases that provide unique functions or multiple missions for the military (Snyder 2001). A comparison of the previous BRAC process and the current changes may shed some light on the differences between the two, and may also indicate whether some level of policy learning has taken place. More than likely, the 2005 BRAC process may be more political than previous closure rounds, while at the same time attempting to appear less so, thus substantiating Koven's argument.

The Koven Thesis

Steve Koven's article about the BRAC process, which appeared in the September-October 1992 edition of *Public Administration Review*, exposed the attempt to revisit the politics-administration dichotomy for the window dressing that it really was. Based on an examination of documents published by the U.S. General Accounting Office, Koven determined that despite appearances, the BRAC process was rife with politics at every step. The legislative intent of Congressman Dick Armey's BRAC process was to allow the military, rather than Congress, to decide which bases should be closed, as had been done in the past. The first three steps in the process focus almost entirely on the bureaucratic elements of closure, except for where the congressional leadership and the president are involved with appointing the eight members of the nonpartisan, or perhaps more accurately, bipartisan, BRAC commission. It is on the early steps that Koven focused his analysis. Briefly, the steps of the BRAC process are:

1. Each branch of the military establishes closure criteria and conducts an internal review, resulting in preliminary lists for closure or realignment, which involves redirecting a mission from one facility to another
2. The defense secretary compiles these lists into an overall preliminary closure and realignment list, then forwards it on to the independent commission
3. The BRAC commission takes the preliminary lists under advisement, conducts regional hearings and constructs the final "hit list." It has extensive powers to reject the preliminary list and start from scratch, or to revise prior BRAC rulings
4. The president has forty-five days to either accept or reject the entire list, and
5. The Congress has forty-five days to either accept or reject the entire list (Koven 1992).

The first step of the BRAC process is where each branch of the military establishes ostensibly objective closure criteria and undertakes an internal audit in order to make a determination as to which installations are excess capacity. Criteria are supposed to include the military value of an installation, potential cost savings, and the impact on both military and civilian communities (Koven 1992, 528). However, there are numerous discrepancies between each branch of the service. The Army appears to have the most systematic approach, assigning an interval-level score for each of the relevant criteria. The system used by the Air Force is similar, but utilizes an ordinal ranking process for exceeding, meeting, or failure to meet minimum standards. This system was augmented with qualitative information in the form of documentation and opinions of the executive group of high-ranking officers that oversaw the Air Force internal review. GAO was most critical of the process used by the Navy to determine which of its and the Marines Corps's bases should be downsized. The Navy also faced the most organized opposition to its process by members of Congress, who argued that the Navy had used questionable criteria to make important decisions affecting their districts (Koven 1992, 528-529). Koven concludes that the military was largely successful in looking out for its organizational priorities, despite the presence of politics inherent in the process. Rather than a Wilsonian separation of politics and administration, the military was forced to walk a tightrope between the need to maintain its institutional interests and long-term political considerations in Congress (Koven 1992).

But evidence in this book and from other sources shows that in terms of tangible economic outcomes, communities may actually benefit more from closures and subsequent conversion to civilian use than from maintaining active-duty installations, provided it is done correctly (CBO 1996 and 2002; Dardia et al. 1996b; Sorenson 1998).

Policy Learning

David Sorenson (1998) evaluates the base realignment, closure, and conversion processes over time, and concludes that all levels of government have learned lessons from each BRAC round about how to streamline these processes. He concludes that base closures are not necessarily bad things, if governments can deal with the transitions effectively. Other research would seem to substantiate this view. Studies conducted by the RAND Corporation in California communities affected by base closure indicate that while unemployment and commerce were hurt in areas immediately adjacent to installations following closure decisions, the effects typically were not long-lasting or widespread (Dardia et al. 1996b). This is also keeping in mind that California was hit particularly hard by the recession of the early 1990s and by numerous base closures and realignments. While defense cutbacks are generally blamed for California's economic crisis, it was the downsizing, retrenchment, and merger of defense and aerospace corporations that had the greater negative effect, rather than base closures (Dardia et al. 1996b).

The Defense Department, which as we have seen has extensive redevelopment programs and grant opportunities for communities affected by base closure, has maintained that the pain of readjustment is not as bad as people fear. By 1998, base reuse had created 45,000 new civilian jobs and garnered 1,000 new tenants on former base properties. For bases that had been closed for at least two years or more, roughly 75 percent of lost civilian jobs had been replaced. Furthermore, the Pentagon demonstrated that in many communities the number of jobs created at former installations exceeds the number of civilians employed while the facility was active duty (DOD 1998). Economic Development Conveyances of former military property, interim leases for new tenants, grants from the Economic Development Agency, Federal Aviation Administration, Office of Economic Adjustment, and remediation of environmental problems are just some of the services provided by the DOD to help with the adjustment process.

Kenneth Mayer and Anne Khademian (1996) are correct that defense is a more visible policy area than academics have given credit, but the nature of the allocational defense politics involved is quite complex, and differs from issue to issue. While there are politics associated with base conversion, they are typically not of the rational choice variety. The statistical analysis and case studies of base conversion efforts provided indicate that inter-local administrative discourse, grant writing and long-range strategic planning, which focus on matching normative values with developmental resources, are more relevant than zero-sum politics. In fact, as chapter V demonstrates, those cases where communities do attempt to play a rational choice game with one another often find themselves losing everything pertinent to making base conversion a success—such as financial resources and community development—to litigation. For base conversion politics, democratic administrative theory, as opposed to the ideas of the Wilsonian dichotomy, or rational choice, is more applicable, though both of these may have their merits in the closure and procurement processes.

BRAC 2005

The Pentagon, under the direction of President Clinton's defense secretary, former Republican Senator William Cohen from Maine, attempted to get congressional approval for more base closures in 1999 and 2001 (GAO 1997), but was unable to get additional BRAC rounds passed by the House. Perhaps congressmen were acting in their rational self-interests, or perhaps it was a partisan stab at a Democratic president by a Republican Congress looking for an excuse to thwart the administration's agenda. Whatever the reason, the Defense Department was forced to wait for the inauguration of a new president for its next downsizing request. The second Bush administration and its new secretary of defense, Donald Rumsfeld, himself in his second stint at the Pentagon, conducted a defense review that, though mostly kept behind closed doors, again requested another BRAC.

According to observers, Rumsfeld's personal style seemed to alienate many of the key players in the review process, and the Bush administration was in danger of losing its request for additional BRACs (Ricks 2001; Schrader 2001). However, the events of September 11, 2001, forced national defense to the top of the policy agenda. Rumsfeld's *Quadrennial Defense Review* (DOD 2001b) released a couple weeks after the attack on the United States mirrored the arguments of base closure proponents in Congress. Representative Vic Snyder (D-AR) and Senator John McCain (R-AZ) maintain that the U.S. military's resources are maldistributed and that the country is ill prepared to fight nonstate or asymmetrical threats to our national security. Though Snyder, Rumsfeld, and President Bush wanted a BRAC for 2003, and McCain wanted two—in 2003 and 2005—they will have to settle for a single closure round in 2005. This was mandated by the conference committee's final version of the 2003 Defense Budget.

Less Is More

Why close more bases during wartime? If there was ever a time when excess military capacity was needed, surely it would be following the worst attack on U.S. soil in American history, right? Actually, the reality is otherwise. In attempting to fight a war against nonstate actors who seek refuge throughout the world in a network of failed states, the appropriate response is not mass mobilization, but special operations conducted by elite forces. Such postmodern warfare is characterized by rapid overseas deployment of Deltas, Rangers, Marines and other special operations units supported by extensive air and sea power. This is an entirely different kind of warfare from what the United States experienced for most of the twentieth century. Both Representative Snyder (2001) and Defense Secretary Rumsfeld in his *Quadrennial Defense Review* (DOD 2001b) make such claims with regard to postmodern warfare and the counterintuitive need for defense downsizing.

First, they argue that closing bases saves money, which could be better spent fighting terrorists. Savings from prior BRAC rounds are close to an estimated $6 billion annually (CBO 1996). Savings from an additional round would save another $3 billion annually (Snyder 2001). That means $9 billion that can be spent on counterinsurgency long after the present exuberance over increased defense spending has waned. If the 1980s are any indication, political support for defense spending will decline after the immediate crisis has passed. Eventually voters will balk at $400 billion annual defense budgets once pressures on the federal deficit—recently reported to be a record $455 billion for 2003 and projected to be more than $500 billion for 2004—and threats to politically sacred cows such as Social Security and Medicare begin to mount.

Second, the pressure of having to maintain unneeded infrastructure, while ordering greater numbers of deployments with fewer troops, causes our forces to be overworked and stressed out (Snyder 2001). The United States has an increasing number of obligations overseas for combat, peacekeeping, and humanitarian missions. Financial compensation for these troops has also fallen, resulting in pressure to increase pay and benefits to active duty personnel, and their dependents. Failure to do so has resulted in retention problems, especially among Air Force pilots, which affects overall readiness. The costs of such activities are not just financial. They also place a burden on those troops left behind to maintain base facilities and carry out multiple tasks, not to mention the burden on their families.

Third, strategic considerations require that the individual services cooperate more in joint operations. To minimize deployment confusion and friendly-fire casualties, more joint operation training facilities are needed (Snyder 2001; DOD 2001a). Utilizing facilities with joint service operations and multiple missions is crucial if the military is to carry out seamless operations that are not bogged down by interservice rivalry, or intraservice miscommunications between special operations forces and regular troops. This could help avert battlefield disasters such as those portrayed by the tragic story *Black Hawk Down* in

Mogadishu, Somalia, in 1993 (Bowden 1999). In fact, the trend I observed anecdotally from previous BRAC rounds seemed to be that those bases that had multiple missions, and especially those with joint service operations, were the safest from closure or realignment. A reason for this might be that basic personnel requirements are more economical at facilities with multiple missions, thus enhancing their military value. The trend toward joint operations would be facilitated by realignments in another BRAC round (Snyder 2001).

Base Closure Process Reform

While it had the appearance of neutral objectivity, the BRAC process is generally believed to have been too political. Following the 1995 BRAC, which was the final round enabled by the 1990 legislation, the U.S. General Accounting Office issued a report that analyzed the policy lessons learned from the BRAC process. Since only about half of the unneeded facilities were closed, there must have been something wrong with the system. The 1995 round, which was supposed to have been the "mother of all base closures," was nothing of the sort. GAO concluded that while the closure criteria were sound, they could be strengthened. It also argued that the defense secretary should be a more active player for the purpose of coordination between the different branches of the service. In particular, GAO maintains that there is still too much redundancy between the services when it comes to expensive endeavors such as maintenance and logistics depots. It also argues that cost savings have been underestimated (GAO 1997). While the savings might be obscured by costs associated with the military drawdown and environmental cleanup, the closure costs are up front, and the costs of environmental restoration would have to be paid regardless of whether the facilities were closed or not (GAO 1997).

As a result of these criticisms and new strategic requirements, several reforms have been included in the closure process for 2005. Generally, they are intended to increase the power of the military bureaucracies and the defense secretary, while somewhat limiting the powers of the commission and the president. Again, just as the previous process was designed, these reforms are supposed to separate politics from administrative concerns. However, a closer look will indicate that again, politics has not been eliminated, but simply displaced. The first time it was displaced from Congress to the BRAC Commission, this time it will be displaced from the commission and the president to the Pentagon bureaucracy.

Reforms to the closure and realignment process are threefold. First, there is a pre-BRAC review conducted by the Pentagon and defense secretary prior to step 1 to determine which bases should be protected from the closure threat. Second, the ability of the commission to remove facilities from the closure and realignment lists created by the military is severely curtailed. And third, privatization-in-place is only permissible as a reuse solution if it is expressly approved in advance by the commission (DOD 2001b). There is a rationale behind each of these changes.

The pre-BRAC step, which precedes the first two steps taken by the branches of the military and Defense Secretary in the usual process, is to determine solely based on military value, those facilities which should be protected from closure and realignment (DOD 2001b). The emphasis on military value is most likely intended to let the Pentagon leadership determine for itself, based on its own perception of needs as defined in the *Quadrennial Defense Review* (DOD 2001b), how to transform its varying missions to meet new threats without interference from locally employed defense consultants. The Pentagon would prefer to rely more on newer technology, unmanned aircraft, and joint and special operations. It would be more than happy to reduce the use of outdated expensive and high maintenance equipment, so as to create a leaner, meaner military (Hendren 2002).

Congress, on the other hand, due to the collective irrationality built into the institution, has a vested interest in keeping its expensive procurement and maintenance contracts, as well as military construction projects, in the defense budget. By placing roughly half of the bases on an "untouchable" list, the military would seem to be able to mollify much of the congressional opposition to more closures. But what if the bases in states or districts represented by the congressional leadership or heads of the defense committees in Congress are deemed to not have military value by the Pentagon? Or what if the military uses political criteria instead of military value to preserve such installations against their statutory mandate to avoid a showdown with Congress? In either instance, this ostensibly nonpolitical step becomes a hotbed of political controversy between Congress and the military, even before the actual BRAC process begins.

The second change to the process involves the commission itself. The number of commissioners has been increased from eight to nine, but this is relatively minor. What is more significant is that a supermajority of seven is required to reverse a closure or realignment decision on the list forwarded to the commission by the defense secretary (DOD 2001b). Previous BRACs had the authority to completely ignore the military value assessments and preliminary closure lists sent by the defense secretary if they chose to, or even reverse the decisions of previous BRAC rounds. The 1993 BRAC in fact did such things when it redirected the mission of Naval Air Station Dallas to the former Carswell Air Force Base in Fort Worth, Texas, which was right in the middle of the base conversion process mandated by the 1991 BRAC. This was especially unfortunate for the Fort Worth conversion effort, because it already had the Lockheed Corporation on the property as a tenant, and was in the middle of an extensive environmental cleanup. The 1993 realignment complicated matters considerably (Sorenson 1998).

If the 2005 BRAC Commission decides to reject a decision made by one of the services or by the defense secretary, there are strict criteria that must be adhered to. These involve the strategic availability of comparable installations worldwide, consistency with the force-structure plan, selection (military value) criteria, and a consultation with the defense secretary, in either an open or closed forum, prior to the reversal of the decision (DOD 2001b). This would seem to create a procedure where the commission in fact defers to the defense secretary in the event of a disagreement, thus compromising the independent nature of the commission with bureaucratic politics.

The third reform involves removing privatization-in-place as a redevelopment alternative, except in special circumstances. Privatization-in-place is a compromise solution to base closure where the work that had previously been done by the military is shifted to a private contractor, who then largely manages the facility and hires back the people who were displaced by the closure. This has been tried at former installations that employed large numbers of civilian employees, where closure or realignment was politically unpalatable due to the large numbers of jobs lost. While privatization may sound like a good idea to alleviate the effects of base closure or realignment, it is not. Two General Accounting Office studies (GAO 1999a and b) have concluded that privatization-in-place largely failed to replace any of the lost jobs and income resulting from BRAC action, despite spending millions in grant money.

Privatization-in-place was also seen as way to subvert the spirit of BRAC with partisan politics by President Clinton. When the Air Logistics Centers (air maintenance depots) at Kelly AFB in San Antonio, Texas, and McClellan AFB in Sacramento, California, were placed on the 1995 BRAC hit list, the Clinton administration quickly tried to limit the political fallout in these two electorally important states. By offering privatization-in-place as a compromise solution, along with millions of dollars in redevelopment grant money, it felt the damage could be minimized (Sorenson 1998). The only problem is, it does not matter who is doing the work if there is no demand for the product or service being produced, in these two cases, military aircraft maintenance, which was no longer required at previous levels of operation (GAO 1999a and b)—hence, the determination of both facilities to be excess capacity in the first place. As a result, the power of future presidents to subvert closures or realignments with privatization has been curtailed, unless it is approved in advance by the commission.

Observations and Comments

While there seems to have been some level of policy learning in the intervening years between the previous BRAC rounds and the proposed new one in 2005, it has yet to be sufficiently demonstrated that politics can be removed from the process. As such, the Koven thesis holds. Politics cannot be separated from administration by the use of independent commissions. Instead, politics is displaced and takes many forms.

The BRAC 2005 round promises to emphasize bureaucratic politics at virtually every step. However, congressional rationality may manifest itself during and after the pre-BRAC step when the "protection list" is created, as well as during the public BRAC commission hearings, which provide a venue for angry congressmen to stand up for their constituents. The actual powers of the commission will be reduced, to prevent arbitrary or inconsistent actions, but this may have unintended consequences. The independence of the commission may be compromised if it is forced to bend to the will of the defense secretary—making the process appear more political, not less—which is what previous BRACs were accused of doing. This could make the problem worse.

However, by virtually eliminating privatization-in-place as a redevelopment solution, it seems that the power of the president to subvert the process has been reduced. This may prove to be successful. But an enterprising president (whomever it is in 2005) may find a different way to manipulate the process, perhaps by working with the congressional leadership in attempting to stack the deck with BRAC commissioners who favor privatization as a solution to "the evils of bureaucratic inefficiency." The hydra created by the constitutional checks and balances will continue to raise its ugly multiple heads, in spite of attempts to separate politics from administration. But paradoxically, the separation of powers may also have the ability to save the process from itself, depending on the nature of the institutional politics involved.

Policy Recommendations

What might make more sense is if everyone simply acknowledged the political interests involved and acted accordingly. The Pentagon prefers to make its own assessments regarding domestic defense infrastructure and readiness. It should be allowed to do so. However, it is accountable to congressional oversight and the budget process, so it must operate within those constraints. Congress has its own collective dilemma to solve, while the president has electoral concerns as well as having to maintain the big picture regarding overall defense and foreign policy capabilities.

Here is what I propose: A three-step closure and realignment process would be less time consuming, less costly, and probably less controversial, because the politics would be up front. The first step would require an internal review on the military value of installations by the Pentagon, resulting in a closure and realignment hit list. During this step, the defense secretary would conduct reviews of each branch of the service based on clear and consistent criteria, hold public hearings and site visits, and formulate the overall strategy. The second step would require approval or disapproval by the president of the entire list, as is the current practice. The third step would require the approval or disapproval by the Congress of the entire list, which is also the current practice. Privatization or other expensive policy boondoggles could be outlawed by statute.

This plan would eliminate (1) formulating the pre-BRAC "protection list," (2) inconsistent closure criteria by each of the services, (3) the existence, and hence, the budget of the BRAC Commission itself, (4) Congress's collective dilemma, and (5) unneeded expensive compromises, which subvert the legislative intent of the process. The benefits include a reduction in interservice rivalry, the elimination of the façade of the political neutrality that fools no one but the voters, and a resolution of the collective irrationality problem. Perhaps this could be done automatically as part of the Quadrennial Defense Review required by Congress, which would make either the closing and realigning, keeping, or adding of facilities a regular activity depending on the strategic requirements of the day. While the defense secretary would come under a great deal of fire from both the military and Congress, he already does anyway. This solution would not separate politics from administration, but displace it from Congress onto the office of the defense secretary, which is essentially how it is going to be done in 2005, but without the confusion created by the independent commission. The big difference is that my solution involves fewer steps and recognizes the process for the contentious, redistributive politics that it is.

Key Pieces of Legislation

HR 1235. *Encroachment on Military Bases Prevention Act*. Submitted by Reps. Elton Gallegly (R-CA), and Jim Gibbons (R-NV), March 12, 2003. http://thomas.loc.gov.

HR 1835. *National Security Readiness Act of 2003*. Submitted by Reps. Elton Gallegly (R-CA), Richard Pombo (R-CA), Jim Gibbons (R-NV), and Bill Young (R-FL), April 29, 2003. http://thomas.loc.gov.

SA 722. Amendment to the *Defense Authorization Act of 2004* (S 1050). Submitted by Sens. Frank Lautenberg (D-NJ), and Jim Jeffords (I-VT), May 28, 2003. http://thomas.loc.gov.

References

Aharoni, Yair. *Markets, Planning and Development.* Cambridge, MA: Ballinger Publishing Co., 1977.

Air Force Times. "Health Effects of Closed Facilities to be Studied." October 19, 1998): 14.

Allison, Graham. *Essence of Decision: Explaining the Cuban Missile Crisis.* Glenview, IL: Scott, Foresman and Co., 1971.

Amenta, Edwin, and Theda Skocpol. "Taking Exception: Explaining the Distinctiveness of American Public Policies in the Last Century." In *The Comparative History of Public Policy*, edited by F. G. Castles (New York: Oxford University Press, 1989).

Anderson, Terry, and Donald Leal. *Free Market Environmentalism.* San Francisco, CA: Westview Press, 1991.

Anderson, Terry, ed. *NAFTA and the Environment.* San Francisco, CA: West view Press, 1993.

Argyros, George. "Voters Decided Means to Plan El Toro" *Los Angeles Times* (January 8, 1995): C10.

Associated Press. "Books Replace Guns, Jeeps at Army Base Converted to College." *Lubbock Avalanche-Journal*, August 29, 1995, B7.

———. "Ex-Army bases can turn profit for communities: Remaking sites for civilians spurring economies" *Dallas Morning News* (January 17, 1999): 42A.

———. "Clean-up Battle is on over Live Munitions," *Orange County Register*, December 26, 2001.

Bachach, P. and and Baratz, M. "The Two Faces of Power." *American Political Science Review* 56, (1962): 947-952.

Baum, Dan. "The Battle of Fort Ord." *Los Angeles Times*, April 27, 2003, B7.

Baumgartner, Frank, and Bryan Jones. *Agendas and Instability in American Politics*. Chicago: University of Chicago Press, 1993.

Behn, Robert D. *Rethinking Democratic Accountability*. Washington, DC: Brookings Institution Press, 2002.

Bennett, Noel. "Fort Chaffee Army Reservation." (Dallas, TX: EPA Region 6, 2002).

Benveniste, Guy. *Mastering the Politics of Planning*. San Francisco, CA: Jossey-Bass, 1989.

Berg, T. "Busy Days in El Toro Fight" *Orange County Register* (February 8, 1995): A1.

Berstrom Air Force Base Conversion Agency (AFBCA). "Installation History and Restoration Program." (Austin, TX: Bergstrom Redevelopment Author ity, 1995).

Bosso, Christopher, and Deborah Lynn Guber. "The Boundaries and Contours of American Environmental Activism." In *Environmental Policy: New Directions for the Twenty-first Century*, 5th ed., edited by Norman Vig and Michael Kraft. Washington, DC: CQ Press, 2003.

Bowden, Mark. *Black Hawk Down*. New York: Penguin Books, 1999.

Buchholz, Rogene, Alfred Marcus and James Post. *Managing Environmental Issues: A Casebook*. Englewood Cliffs, NJ: Prentice Hall, 1992.

Bullard, Robert D. "Anatomy of Environmental Racism and the Environmental Justice Movement." In *Debating the Earth: The Environmental Politics Reader* edited by John S. Dryzek and David Scholsberg. New York: Oxford University Press, 1998.

Carman, Christopher J. "Brokering Interests: The Use of Market-Based Policies by Interest Groups." *Social Science Quarterly* 83, no. 1 (March 2002): 137-155.

Cecil Field Development Commission. *NAS Cecil Field Final Base Reuse Plan*. Jacksonville, FL: Arthur Anderson & Co., February 1996.

Center for Public Service (CPS). *Conversion Update* newsletter. Lubbock, TX: Texas Tech University, July 13, 1995.

Cochran, Clarke, Lawrence Mayer, T. R. Carr, and N. Joseph Cayer. *American Public Policy: An Introduction*, 6th ed. New York: Worth Publishers, 1999.

Congressional Budget Office (CBO). *CBO Papers: Closing Military Bases, an Interim Assessment*. Washington, DC: Congressional Budget Office, December 1996.

Cook, Brian J. "The Politics of Market-Based Environmental Regulation: Continuity and Change in Air Pollution Control Policy Conflict." *Social Science Quarterly* 83, no. 1 (March 2002): 156-166.

Costain, W. Douglas, and James Lester. "The Evolution of Environmentalism." In *Environmental Politics and Policy*, 2d ed., edited by James Lester. Durham, NC: Duke University Press, 1995.

Curry, Tom. "What Will become of the Budget?" http://www/msnbc.com/news/697914.asp?pne=11947. (accessed February 4, 2002).

Dardia, Michael, McCarthy, Kevin, Malkin, Jesse, and George Vernez. *Defense Cutbacks: Effects on California's Communities, Firms, and Workers—Executive Summary*. Santa Monica, CA: Rand Corporation, 1996a.

———. *The Effects of Military Base Closures on Local Communities*. Santa Monica, CA: Rand Corporation, 1996b.

Davis, Charles E. *The Politics of Hazardous Waste*. Englewood Cliffs, NJ: Prentice Hall, 1993.

Department of Defense (DOD). *Report of the Department of Defense on Base Realignment and Closure*. (April 1998) www.defenselink.mil (accessed April 1998).

———. *Economic Renewal: Community Reuse of Former Military Bases*. (April 21, 1999) www.defenselink.mil/pubs/reuse (accessed May 1999).

———. *Quadrennial Defense Review*. (September 2001a). www.defenselink.mil (accessed September 21, 2001).

———. *Efficient Facilities Initiative*. (December 2001b). www.defenselink.mil (accessed January 2002).

Devens Enterprise Commission. *Sustainability Indicator Report*. (March 2002) http://www.devensec.com/sustainreport.html (accessed May 25, 2003).

Domahidy, Mary R. and James F. Gilsinan. "The Back Stage Is Not the Back Room: How Spatial Arrangements Affect the Administration of Public Affairs." *Public Administration Review* 52, no. 6 (1992): 67-80.

Downs, Anthony. "Up and Down with Ecology: The 'Issue Attention Cycle.'" *Public Interest* 28 (1972): 38-50.

Dryzek, John S., and David Scholsberg, eds. *Debating the Earth: The Environmental Politics Reader*. New York: Oxford University Press, 1998.

Dunlap, Riley E. "Public Opinion and Environmental Policy." In *Environmental Politics and Policy*, 2d ed., edited by James Lester. Durham, NC: Duke University Press, 1995.

Durbin, Harry. *The Military and the Environment: A Case Study of Reese Air Force Base*. Undergraduate Senior Thesis, General Studies, Texas Tech University, 1993.

Dye, Thomas, and Harmon Zeigler. *The Irony of Democracy*, Millenium ed. Fort Worth, TX: Harcourt-Brace Publishers, 2000.

Easton, David. *The Political System: An Inquiry into the State of Political Science*. New York: Knopf, 1953.

Eckles, Paul. "ETRPA Millenium Plan." El Toro Reuse Planning Authority bro chure, December, 2000.

Elmore, Richard. "Forward and Backward Mapping: Reversible Logic in the Analysis of Public Policy." In *Policy Implementation in Federal and Unitary Systems: Questions of Analysis and Design*. Edited by Kenneth Hanf and Theo A. J. Toonen. Dordrecht: Martinus Nijhoff, 1985.

Environmental Protection Agency (EPA). *BRAC Success Story: Bergstrom Air Force Base* (February 1998). http://www.epa.gov/swerffrr (accessed Febru-Ary 1998).

———. *National Priorities List (NPL) Fact Sheet*. (January 30, 2001) http://yosemite1.epa.gov./rl/npl_pad.nsf (accessed February 2003).

———. *El Toro Marine Corps Air Station*. (November 26, 2002) http://yosemite1.epa.gov./rl/npl_pad.nsf (accessed May 29, 2003).

———. *Air Force Plant 4 (Lockheed-Martin) Texas*. (February 4, 2003). http://yosemite1.epa.gov./rl/npl_pad.nsf (accessed February 2003).

Farmer, David. *The Language of Public Administration*. Tuscaloosa, AL: University of Alabama Press, 1995.

Federal Facilities Policy Group. *Improving Federal Facilities Cleanup*. (October 1995).

Federal Register. *Rules and Regulations*, 59, no. 26 (October 26, 1994): 740-741.

Ferris, James, and Elizabeth Graddy. "Contracting Out: For What? With Whom?" *Public Administration Review* (July/August 1986): 332-344.

Field, Barry. *Environmental Economics: An Introduction*. (New York: McGraw-Hill, Inc., 1994).

Fort Chaffee Redevelopment Authority (Trust). *Fort Chaffee Comprehensive Reuse Plan*. Fort Smith, AR: City of Fort Smith, 1998.

Fort Smith, City of. *Agreement to Adjust Municipal Boundaries*. Fort Smith, AR: September 2002.

Fox, Charles J., and Hugh T. Miller. *Postmodern Public Administration: Toward Discourse*. Newbury Park, CA: Sage Publications, 1995.

Freeman, A. Myrick, III. *The Economic Approach to Environmental Policy: The Selected Essays of A. Myrick Freeman, III*. Northampton, MA: Edward Elgar, 1998.

Frey, Bruno. *Democratic Economic Policy*. New York: St. Martin's Press, 1983.

Fritz, John. "Cecil Touted for Senior Living." *Florida Times-Union*. February 8, 1998, A2.

Fulton, William, and Paul Shigley. "The Greening of the Brown" *Governing* 14, no. 3 (December 2000): 31-34.

General Accounting Office (GAO). *Military Bases: Lessons Learned from Prior Closure Rounds*. Washington, DC: Government Printing Office, July 1997. GAO/NSAID-97-151.

———. *Superfund: Times to Complete Site Listing and Cleanup*. Washington, DC: Government Printing Office, February 1998. GAO/RCED: 74.

———. *Public Private Competitions: Reasonable Processes Used for Sacramento Depot Maintenance Award*. Washington, DC: Government Printing Office, May 1999a. GAO/NSAID-99-124.

———. *Public Private Competitions: Reasonable Processes Used for San Antonio Engine Depot Maintenance Award*. Washington, DC: Government Printing Office, May 1999b. GAO/NSAID-99-124.

———. *Military Base Closures: Progress in Completing Past Actions from Prior Realignments and Closures*. Washington, DC: Government Printing Office, April 2002. GAO-02-433.

Gibson, Clark C., Fabrice E. Lehoucq, and John T. Williams. "Does Privatization Protect Natural Resources? Property Rights and Forests in Guatemala." *Social Science Quarterly* 83, no. 1 (March 2002): 206-225.

Glassberg, Andrew. "Intergovernmental Relations and Base Closing." *Publius* 25, no. 3 (1995): 87-98.

———. 1996. "The Economic Impact of Military Base Closings: Issues in Measurement." Paper presented at the Southwest Social Science Meeting. New Orleans, LA: March 28, 1996.

Gooden, Robert E. "Selling Environmental Indulgences." In *Debating the Earth: The Environmental Politics Reader*. Edited by John S. Dryzek and David Schlosberg. New York: Oxford University Press, 1998.

Goodsell, Charles T. *The Social Meaning of Civic Space: Studying Political Authority through Architecture*. Lawrence, KS: University Press of Kansas, 1988.

Gottschneider, Richard, Jimmy Hicks, and Jeffrey Donohoe. "Acquiring Property at a Former Military Base: The Process and the Law" *Real Estate Issues* 19 (1994): 1-28.

Grodzins, Morton. "The American System." In *Classics of Public Administra-*

tion, 4th ed., edited by Jay Shafritz and Albert Hyde. Fort Worth, TX: Harcourt Brace Publishers, 1997.

Guskind, R. "Picking Up the Pieces" *Planning*, no. 1 (Spring 1990): 32-33.

Hansen, Kenneth N.. "Pro-Active Environmental Policy: A Case Study of the City of Lubbock" *Texas Journal of Political Studies* 17, no. 2 (Spring/ Summer 1995): 51-65.

———. *Discourse and Complex Implementation: Military Base Conversions in Texas*. Ph.D. diss. Texas Tech University, 1996.

———. "Identifying Facets of Democratic Administration: The Empirical Referents of Discourse" *Administration & Society* 30, no. 4 (September 1998): 443-461.

———. "The Discourse Heresies: Utilizing the Empirical Referents to Evaluate Military Base Transitions" *Administration & Society* 35, no. 3 (July 2003): 315-333.

Hansen, Kenneth N., and Tracy A. Skopek. "The Loss of Military Bases in Texas: Opportunities and Challenges." In *Texas Politics and Policy: A Workbook*, edited by Mark Somma. New York: Simon & Schuster Publishing, 1996.

Hansen, Kenneth N., Skopek, Tracy A., and Mark S. Somma. "The Fundamentals of Local Implementation in Military Base Redevelopment Efforts." *The American Review of Public Administration* 27, no. 4 (December 1997): 377-397.

Hansen, Kenneth N., and Daniel T. Krejci. "Rethinking Neoinstitutional Interac tion: Municipal Arena-Specific Strategies and the Base Closure Process" *Administration & Society* 32, no. 2 (May 2000): 166-182.

Heclo, Hugh. *A Government of Strangers*. Washington, DC: The Brookings Institution, 1977.

Hedge, David M. and Michael J. Schicchitano. "Devolving Regulatory Authority: The Federal and State Response" *Policy Studies Review* 11, no. 1 (Spring 1992): 81-90.

Hedge, David M. *Governance and the Changing American States*. Boulder, CO: Westview Press, 1998.

Hendren, John. "For Some, $48 Billion Defense Boost is a Bust." *Los Angeles Times*, February 1, 2002. http://latimes.com/la-020102defbud.story (ac cessed February 1, 2002).

Herson, Lawrence, and John Bolland. *The Urban Web: Politics, Policy and Theory*. New York: Burnham, Inc., 1998.

Hill, Kim Quaile, and Kenneth R. Mladenka. *Texas Politics and Government: A Concise Survey*, 2d ed. Needham Heights, MA: Allyn & Bacon, 1997.

Hird, John A. *Superfund: The Political Economy of Risk*. Baltimore, MD: Johns Hopkins University Press, 1994.

Hjern, Benny. "Implementation Research—The Link Gone Missing." *Journal of Public Policy* 2, no. 3 (1983): 301-308.

Hughes, Dave. "Fort Chaffee: Director of Panel over Fort Resigns." *Arkansas Democrat-Gazette*, (September 1, 2001), 1B-2B.

Huntington, Samuel P. *The Common Defense: Strategic Programs in National Politics*. New York: Columbia University Press, 1961.

Jacob, Gerald. "The Legacy of the Pentagon: The Myth of the Peace Dividend." In *The Pentagon and the Cities*, edited by Andrew Kirby. Newbury Park, CA: Sage Publications, 1992.

Kettl, Donald F. *Reinventing Government? Appraising the National Performance Review*. Washington, DC: The Brookings Institution, 1994.

King, Gary, Ronald Keohane, and Sydney Verba. *Designing Social Inquiry: Scientific Inference in Qualitative Research*. Princeton, NJ: Princeton University Press, 1994.

Kingdon, John. *Agendas, Alternatives and Public Policies*, 2d ed. New York: Longman, 1984.

Kirby, Andrew, ed. *The Pentagon and the Cities*. Newbury Park, CA: Sage Publications, 1992.

Knott, Jack H., and Gary J. Miller. *Reforming Bureaucracy: The Politics of Institutional Choice*. Englewood Cliffs, NJ: Prentice-Hall, Inc., 1987.

Koven, Steven. "Base Closings and the Politics-Administration Dichotomy Revisited." *Public Administration Review* 52, no. 5 (1992): 526-533.

Kraft, Michael. "Congress and Environmental Policy." In *Environmental Poli-*

tics and Policy, 2d ed., edited by James Lester. Durham, NC: Duke University Press, 1995.

———. *Environmental Policy and Politics*. New York: Longman, 1996.

Lee, Kai N. "The Columbia River Basin: Experimenting with Sustainability." *Environment* 31, no. 6 (1989): 6-11, 30-33.

Lester, James. "Federalism and State Environmental Policy." In *Environmental Politics and Policy*, 2d ed., edited by James Lester. Durham, NC: Duke University Press, 1995.

Lipski, Michael. "Toward a Theory of Street-Level Bureaucracy." In *Theoretical Perspectives on Urban Politics*, edited by Willis D. Hawley et al. Englewood Cliffs, NJ: Prentice-Hall, 1976.

Los Angeles Times. "Extensive Pollution Stalls Base Conversions" (May 1, 2000). www.latimes.com (accessed May 30, 2000).

Lowi, Theodore. *The End of Liberalism*. New York: W. W. Norton & Co., 1969.

Maremont, M. "Lessons from the First Base to Go Dark." *Business Week*, April 5, 1993, 25.

Mayer, Kenneth. "Closing Military Bases (finally): Solving Collective Dilemmas through Negotiation." *Legislative Studies Quarterly* 20 (1995): 393-413.

Mayer, Kenneth, and Anne M. Khademian. "Bringing Politics Back in: Defense Policy and the Theoretical Study of Institutions and Processes." *Public Administration Review* 56, no. 2 (1996): 180-190.

Mayer, Virginia. *Local Officials Guide to Defense Economic Adjustment*. (Washington, D.C.: National League of Cities, 1992).

Mazmanian, Daniel A., and Paul A. Sabatier. *Implementation and Public Policy*. New York: University Press of America, 1983.

McSpadden, Lettie. "The Courts and Environmental Policy." In *Environmental Politics and Policy*, 2d ed., edited by James Lester. Durham, NC: Duke University Press, 1995.

McSwite, O. C. "Postmodernism, Public Administration, and the Public Interest." In *Refounding Democratic Public Administration: Modern Paradoxes,*

Postmodern Challenges. Edited by Gary L. Wamsley and James F. Wolf. Thousand Oaks, CA: Sage Publications, 1996.

Meckstroth, Theodore W. "Most Different Systems and Most Similar Systems: A Logic of Comparative Inquiry." *Comparative Political Studies*. 8, no. 2 (1975): 132-157.

Melcher, Bonita H., and Harold Kerzner. *Strategic Planning: Development and Implementation*. Blue Ridge Summit, PA: Tab Books, 1988.

Mennis, Jeremy. "Using Geographic Information Systems to Create and Analyze Statistical Surfaces of Population and Risk for Environmental Justice Analysis." *Social Science Quarterly* 83, no. 1 (March 2002): 281-297.

Mercer, James L. *Public Management in Lean Years*. Westport, CT: Quorum Books, 1992.

Mercer, James L., Susan Woolston, and William Donaldson. *Managing Urban Government Services*. New York: Amacom, 1981.

Milbourne, M., and J. Pasco. "A Boost for El Toro Airport?" *Orange County Register* March 8, 1995, A1.

Miller, Hugh T. "Method: The Tail that Wants to Wag the Dog." *Administration & Society* 30, no. 4 (September 1998): 443-461.

Mintzberg, Henry. *The Rise and Fall of Strategic Planning*. New York: The Free Press, 1994.

Mitchell, William C., and Randy T. Simmons. "Political Pursuit of Private Gain: Environmental Goods." In *Debating the Earth: The Environmental Politics Reader*, edited by John S. Dryzek and David Scholsberg. New York: Oxford University Press, 1998.

Moe, Ronald. "Exploring the Limits of Privatization." In *Classics of Public Administration*, 4th ed., edited by Jay Shafritz and Albert Hyde. Fort Worth, TX: Harcourt Brace, 1997.

Morgan, Anne M. "Bobbi," and James L. Meyer. *Funding for Military Base Reuse*. Alexandria, LA: National Association of Installation Developers, 1995.

Morgan, David, and Robert England. "The Two Faces of Privatization" *Public Administration Review* (November/December, 1988): 979-987.

National Association of Installation Developers (NAID). *NAID News* no. 3 (August-November 1994): 1-5.

National Performance Review (NPR). *From Red Tape to Results: Creating a Government that Works More and Costs Less*. Washington, DC: Govern ment Printing Office, 1993.

Obach, Brian K. "Labor-Environmental Relations: An Analysis of the Relationship between Labor Unions and Environmentalists." *Social Science Quarterly* 83, no. 1 (March 2002): 82-100.

Office of Economic Adjustment (OEA). *Civilian Use of Former Military Bases*. Washington, DC: Department of Defense, 1993.

Oldham, Jennifer. "El Toro Sale Puts Regional Plans in a Fix" *Los Angeles Times* (April 24, 2002): A7.

Osborne, David, and Ted Gaebler. *Reinventing Government: How the Entrepreneurial Spirit Is Transforming the Public Sector.* Reading, MA: Addison-Wesley, 1992.

Pagano, Michael A. and Ann O'M. Bowman. *Cityscapes and Capital*. Baltimore, MD: Johns Hopkins University Press, 1995.

Pasco, Jean O. "Navy Sinks Last Chance of Airport at El Toro" *Los Angeles Times* (April 24, 2002): A1-A7.

Pasco, Jean O., Scott Martelle, and Dan Weikel. "In Jolt to Orange County, Navy to Sell El Toro Land" *Los Angeles Times* (March 7, 2002): A1.

Pastor, Manuel Jr., James L. Sadd, and Rachel Morello-Frosch. "Who's Minding the Kids? Pollution, Public Schools, and Environmental Justice in Los Angeles." *Social Science Quarterly* 83, no. 1 (March 2002): 263-280.

Peters, B. Guy, and Donald J. Savoie. "Managing Incoherence: The Coordination and Empowerment Conundrum." *Public Administration Review* 56, no. 3 (1996): 281-290.

Peterson, Paul E. *City Limits*. Chicago: University of Chicago Press, 1981.

Pike, John. *Fort Devens, Devens Reserve Force Training Area 33, Devens MA.* (September 27, 2002). http://www.globalsecurity.org/military/facility/fort-devens.htm (accessed May 2003).

Pine, Art. "Texas Base's Rebirth Shows Military Holding its Ground" *Los Ange-*

les Times (May 28, 1994): A24-A25.

Pine, John C., Brian D. Marx, and Aruna Lakshmanan. "An Examination of Accidental-Release Scenarios from Chemical Processing Sites: The Relation of Race to Distance." *Social Science Quarterly* 83, no. 1 (March 2002): 317-331.

Polsby, Nelson. *Community Power and Political Theory*, 2d ed. New Haven, CT: Yale University Press, 1980.

Rabe, Barry. *Beyond NIMBY: Hazardous Waste Siting in Canada and the United States*. Washington, DC: Brookings Institution, 1994.

———. "Power to the States: The Promise and Pitfalls of Decentralization." In *Environmental Policy: New Directions for the Twenty-first Century*, 5th ed., edited by Norman Vig and Michael Kraft. Washington, DC: CQ Press, 2003.

Ricks, Thomas E. "Pentagon Deadlocks on Rumsfeld Review" *The Washington Post*, July 15, 2001, 5A.

Ringquist, Evan. *State Level Environmental Policy*. Armonk, NY: M. E. Sharpe, 1993.

———. "Environmental Justice: Normative Concerns, Empirical Evidence, and Government Action." In *Environmental Policy: New Directions for the Twenty-first Century*, 5th ed. Washington, DC: CQ Press, 2003.

Ripley, Randall, and Grace Franklin. *Congress, the Bureaucracy and Public Policy*, 5th ed. Pacific Grove, CA: Brooks/Cole Publishing, 1991.

Rivlin, Alice. *Systematic Thinking for Social Action*. Washington, DC: The Brookings Institution, 1971.

Rohr, John A. *Civil Servants and their Constitutions*. Lawrence, KS: University Press of Kansas, 2002.

Rosenbaum, Walter A. "The Bureaucracy and Environmental Policy." In *Environmental Politics and Policy*, 2d ed., edited by James Lester Durham, NC: Duke University Press, 1995.

Rubenson, David, and John R. Anderson. *California Base Closure: Lessons for DOD's Cleanup Program*. Santa Monica, CA: RAND Corporation, 1995.

Sabatier, Paul A. "An Advocacy Coalition Framework of Policy Change and the Role of Policy Oriented Learning Therein." *Policy Sciences* 21, (Fall 1998): 129-168.

Schlesinger, Arthur. *The Cycles of American History*. Boston: Houghton Mifflin, 1986.

Schrader, Esther. "Pentagon Plan Seeks New Base Shutdowns." *Los Angeles Times*, July 27, 2001. http://latimes.com/news/nationworld/nation/la-072701bases.story (accessed July 27, 2001).

Shogren, Elizabeth. "Colorado Has New Battle to Get Base Cleaned Up." *Los Angeles Times*, April 9, 2003. http://latimes.com/news/la-na-enviro9apr09 (accessed April 9, 2003).

Shutkin, William A. *The Land that Could Be: Environmentalism and Democracy in the Twenty-first Century*. Cambridge, MA: MIT Press, 2000.

Skornek, Carolyn. "Defense Spending Bill with Base Closings Clears Congress" *Arkansas Democrat-Gazette*, August 8, 2001, J2.

Smith, Zachary A. *The Environmental Policy Paradox*, 3d ed. Upper Saddle River, NJ: Prentice Hall, 2000.

Snyder, Vic. "Base Closure, Realignment a Must." *Arkansas Democrat-Gazette*, August 8, 2001, J2.

Somma, Mark S. "Southwestern Water Politics." In *Politics and Process: United States and Texas*. Edited by Nelson Dometrius and Mark Somma. New York: Kendall/Hunt Publishing, 1993.

______. "Local Autonomy and Groundwater District Formation in the High Plains Region of West Texas." *Publius* 24 (Fall 1994): 1-10.

———. "Ideology and Local Solutions to the Tragedy of the Commons: West Texas Groundwater Policy" *Publius* (Spring 1998): 23-42.

Somma, Mark S. and Kenneth N. Hansen. "Core Value Survey: Community, Economy and the Environment in Amarillo, Texas." Sponsored by the Pantex Plant Citizen's Advisory Board. (Lubbock, TX: Center for Public Service, Texas Tech University, 1996).

Sonenshein, Raphael J. *Politics in Black and White: Race and Power in Los Angeles*. Princeton, NJ: Princeton University Press, 1993.

Sorenson, David. *Shutting Down the Cold War: The Politics of Military Base Closure*. New York: St. Martin's Press, 1998.

Stein, Robert M. *Urban Alternatives*. Pittsburgh, PA: University of Pittsburgh Press, 1990.

Steppick, Daniel C. *A Military Mess: CERCLA Liability and the Base Closure and Realignment Act*. Dallas, TX: SMU School of Law, 1994.

Stillman, Richard J. *Preface to Public Administration*. New York: St. Martin's Press, 1991.

Stivers, Cam. "Understanding Social Process: The Key to Democratic Governance." In *Refounding Democratic Public Administration*, edited by Gary Wamsley and James Wolf. Newbury Park, CA: Sage Publications, 1996.

Stone, Clarence, and Heywood T. Sanders, eds. *The Politics of Urban Development*. Lawrence, KS: University of Kansas Press, 1987.

Strickland, Sandy. "Yellow Water Area Quiet...At Least for Now." *Florida Times-Union*, August 25, 1997, A1.

Taquino, Michael, Domenico Parisi, and Duane A. Gill. "Units of Analysis and the Environmental Justice Hypothesis: The Case of Industrial Hog Farms." *Social Science Quarterly* 83, no. 1 (March 2002): 298-316.

Taylor, Frederick W. "Scientific Management" excerpt from testimony before the U.S. House of Representatives, January 25, 1912. In *Classics of Public Administration*, 4th ed., edited by Jay Shafritz, and Albert Hyde. Fort Worth, TX: Harcourt Brace, 1997.

Texas Air Control Board (TACB). *1990-1992 Biennial Report*. Austin, TX: State of Texas, 1992.

Texas Water Commission (TWC). *1992 Annual Report*. Austin, TX: State of Texas, 1992.

Torgerson, Douglas. "Limits of the Administrative Mind: The Problem of Defining Environmental Problems." In *Managing Leviathan: Environmental Politics and the Administrative State*, edited by Robert Paehlke and Douglas Torgerson. Peterborough, ON: Broadview Press, 1990.

U.S. Department of the Navy. *Draft Environmental Impact Statement: Disposal and Reuse of Naval Air Station Cecil Field, Jacksonville, Florida*. April 1997.

Urban, Lloyd. "Texas High Plains." In *Groundwater Exploitation on the High Plains*, edited by David Kromm and Stephen White. Lawrence, KS: University of Kansas Press, 1992.

Vig, Norman, and Michael Kraft, eds. *Environmental Policy: Directions for the Twenty-first Century*, 5th ed. Washington, DC: CQ Press, 2003.

Walker, J. L., Jr. *Mobilizing Interest Groups in America: Patrons, Professions, and Social Movements*. Ann Arbor: University of Michigan Press, 1991.

Wamsley, Gary, and James Wolf. *Refounding Democratic Public Administration*. Newbury Park, CA: Sage Publications, 1996.

Weber, Max. "Bureaucracy." In *Classics of Public Administration*, 4th ed., ed ited by Jay Shafritz and Albert Hyde. Fort Worth, TX: Harcourt Brace College Publishers, 1997.

Wenner, Lettie. "Environmental Policy in the Courts." In *Environmental Policy in the 1990s*, edited by Norman Vig and Michael Kraft. Washington, DC: CQ Press, 1993.

Whicker, Marcia Lynn, and Nicholas Giannatasio. "The Politics of Military Base Closings: A New Theory of Influence." *Public Administration Management Internet Journal(PAMIJ)* 1, no. 1 (1996).

Wilson, James Q. *Bureaucracy*. New York: Basic Books, 1989.

Wilson, Woodrow. "The Study of Administration." In *Classics of Public Ad ministration*, 4th ed., edited by Jay Shafritz and Albert Hyde. Fort Worth, TX: Harcourt Brace College Publishers, 1997.

Wright, Deil S. *Understanding Intergovernmental Relations*, 3d ed. Fort Worth, TX: Harcourt Brace & Co., 1988.

Yin, Robert K. *Case Study Research: Design and Methods*. Newbury Park, CA: Sage Publications, 1989.

Index

About the Author

Kenneth N. Hansen, Ph.D., is a professor of political science at the University of Arkansas, in Fayetteville, Arkansas, and is the advisor for the bachelor's of science in public administration degree program. His work has appeared in *Administration & Society* and *The American Review of Public Administration*, among other national and international publication venues. He continues to research homeland defense organization, military base closure and conversion, and environmental politics. He can be contacted at khansen@uark.edu.

www.ingramcontent.com/pod-product-compliance
Ingram Content Group UK Ltd.
Pitfield, Milton Keynes, MK11 3LW, UK
UKHW041627031026
15587UKWH00001B/113

9 780739 105399